MW00814423

October 1, 2002, with wife in Hawaii.

August 30, 2003, at daughter's wedding

September 28, 2008

Robert K. Su, M.D.

8/10/2010

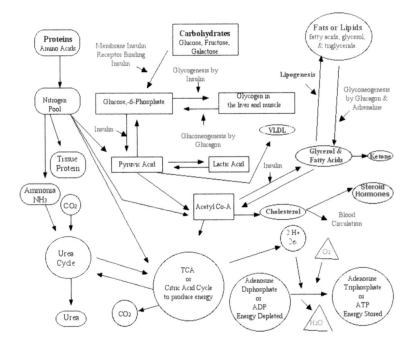

Figure 5. Metabolism of the Nutrients (Page 15)

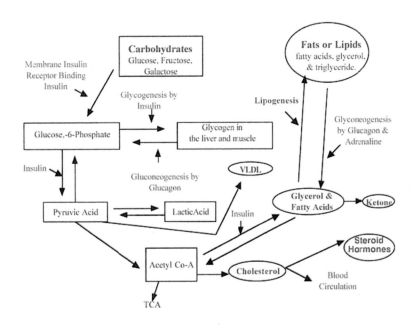

Figure 16. Insulin and Glucose Metabolism (Page 28)

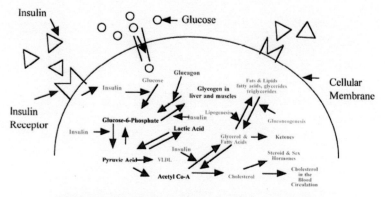

After insulin is coupled with the insulin receptor of the cellular membrane, insulin helps the cell take in glucose; store glucose in the liver and muscles as glycogen; process glucose-6-phosphate to acetyl Co-enzyme A for generating energy in the TCA cycle; and produce fats (fatty acids and glycerols.)

Figure 17. Insulin and Cellular Utilization of Glucose (Page 29)

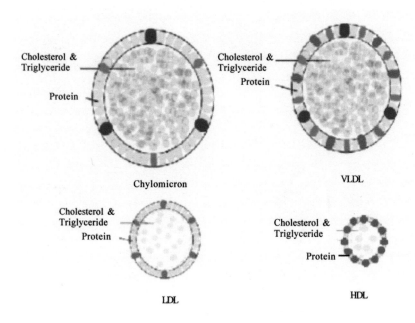

Figure 18. Types of Plasma Lipoproteins (Page 32)

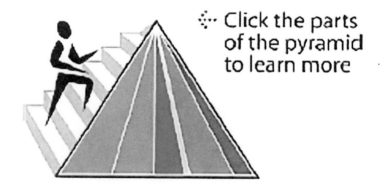

⁘ **Click the parts of the pyramid to learn more**

Color Representation
Chocolate: Grains. **Green**: Vegetables, **Dark Red**: Fruits,
Corn Flower Blue: Milk Group, **Dark Blue**: Meat and Beans.
Source: United States Department of Agriculture
http://www.mypyramid.gov/pyramid/index.html

Figure 23. Inside the Pyramid (Page 57)

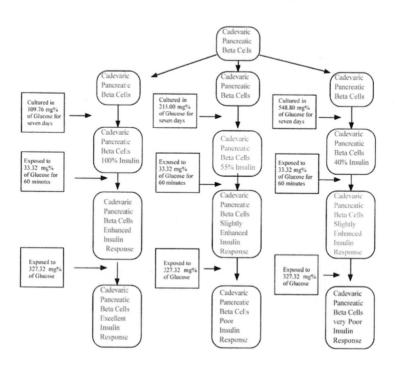

Figure 34. Carbohydrates Abuse Causes Diabetes Mellitus (Page 215)

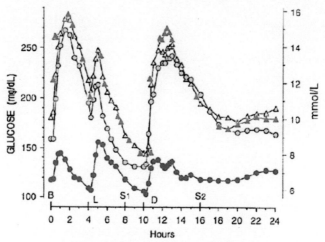

Effect of diet on glucose. Mean plasma glucose concentration before (triangles) and after 5 weeks on control diet (yellow circles: (CHO: fat: protein = 55:30:15)) or 5 weeks on the higher fat diet (blue circles: (20:50:30)). Meals are Breakfast (B), lunch (L) and dinner(D) plus 2 snacks (S1, S2). Data from this article, "Effect of a High-Protein, Low-Carbohydrate Diet on Blood Glucose Control in People With Type 2 Diabetes", modified for the article, "When is a high fat diet not a high fat diet?", that was published in the Nutrition & Metabolism (London). Online 2005 October 17. doi: 10.1186/1743-7075-2-27, by Richard D Feinman

Reprint with permission from Professor Richard D Feinman

(Page 224)

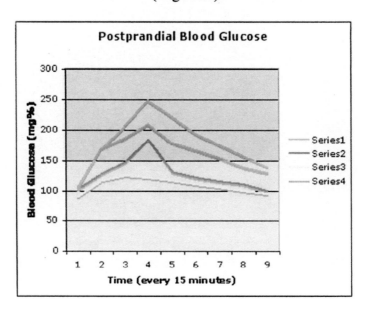

Figure 39. Current Blood Glucose Tests Do Not Tell Us the Whole Story

(Page 365)

With A Physician's Personal Experience,
and A Mountain of Evidence...

CARBOHYDRATES
CAN
KILL

Robert K. Su, M.D.

CARBOHYDRATES
CAN
KILL

ROBERT K. SU, M.D.

Two Harbors Press
Minneapolis, MN

Two Harbors Press
212 3rd Avenue North, Suite 290
Minneapolis, MN 55401
612.455.2293
www.TwoHarborsPress.com

ISBN - 978-1-935097-08-2
ISBN - 1-935097-08-3
LCCN - 2008912209

Book sales for North America and international:
Itasca Books, 3501 Highway 100 South, Suite 220
Minneapolis, MN 55416
Phone: 952.345.4488 (toll free 1.800.901.3480)
Fax: 952.920.0541; email to orders@itascabooks.com

Cover Design and Typeset by Sophie Chi

Printed in the United States of America

October 1, 2002, with wife
in Hawaii, before dieting.

August 30, 2003, at daughter's
wedding in Norfolk

September 28, 2008

Robert K. Su, M.D.

Table Of Contents

Table of Contents

Table of Contents

Table of Contents

Dedication

For the past six years, I have experienced the most memorable time of my life. When I found out I had serious health problems I decided against seeking conventional medical treatment. Instead, I chose to use my own body for experimentation. My decision was a gamble between life and death. For the first six months of the period I felt as though I was sitting on a time bomb -- severe high blood pressure. Not knowing when that bomb might go off, I had become anxious and depressed.

During the entire period, my wife provided patient support and encouragement without a single complaint. Although she was skeptical of my plan at first, she has become a firm believer in what I have accomplished. She had not known much about the impact of carbohydrate consumption on our health. Now she considers the amount of carbohydrates in each food she puts on our dinner table.

Without my wife's unconditional love, I could not have improved my health and the quality of my life. Nor could I have learned so much about the dark side of carbohydrates that I share with you now. I am deeply indebted to her.

With my everlasting love, I wish to dedicate this book to my dear wife, Martha.

Preface

When you first looked at the title of this book, you may have asked yourself, "Can carbohydrates really kill? If they do, why have I had no problem with them?" After you finish reading this book, you will find out why carbohydrates are dangerous to not only our health, but also to the health of future generations.

You probably wonder if this is just another book offering a low carbohydrate diet for weight loss. There are so many dieting books already in the bookstores. There is no need of another one. Obviously, people are tired of hearing about low-carbohydrate diets and weight loss. The editor of the *Virginian Pilot* even titled his editorial article of the January 2, 2005 issue, "Carb-counting: a fad whose time has run." I was very concerned about the impacts on its readers by such an erroneous comment. I responded with a letter to the editor below:

"Do not treat carb-counting a fad. (01/02/2005 Editorial: Carb-counting: a fad whose time has run.) The dietary recommendation with emphasis on a high-carb and low-fat diet for the last several decades deserves a serious second look. Despite the blame on the sedentary lifestyle and overeating, the high-carb and low-fat diet is likely responsible for the increase of cases of cardiovascular and other catastrophic disorders.............."

I have been a medical doctor for 37 years. I have watched the increase in health problems brought on by obesity. Unfortunately, I failed to pay attention to my own increasing weight, because I thought I was eating a healthy, balanced diet. I did not begin to take care of my cardiovascular system until June 9, 2002, when I accidentally found out my blood pressure was dangerously high. I did not want to count pills for the rest of my life so I decided to use my own body for research. I was going to see if I could alter this terrible situation and save my life, without pills. In the meantime, I began to delve into medical journals to identify the real culprit responsible for my problems. As of this writing, I have reviewed more than 1,200 articles and I am still counting. Although I did not necessarily read all articles word for word, I tried to understand and take note of their findings.

After observing improvement in my health, I continued to record my progress. I believe that it is the time for me to share my findings about the dark side of sweets or carbohydrates. I hope I can help you improve your health.

I begin with a chapter titled Nutrition and Health, in which I give some background in food intake and the makeup of the human body. In hope of providing a clearer understanding in explanations to follow, I have included a glossary with graphs in the first chapter. The glossary list should help you clarify some of the terms as you read on.

I also want to discuss ways, in which weight gain and obesity can impact your health. You will quickly see that the body works as a unit, with each and every part affected by diet. If you can cut extra weight with a good diet, you can become healthy again, and live longer without facing so many illnesses. I am going to show you why today's popular dietary recommendations based on a food pyramid are responsible for the increase in the numbers of people suffering ailments related to weight gain.

Next, I will describe a six-year period in my own life, during which I improved my health through acting on findings from my experiments. I detail some of the changes in my physical condition and health to illustrate how I struggled during these years but refused to give up the hope of reclaiming my good health. Perhaps, in reading about my experiences, you will gain insight for facing your own similar hurdles.

I shall also share with you my notes about the findings of 135 out of a reading list of 1,163 articles. For your reference, the reading list is posted online at www.carbohydratescankill. com. The articles included in this book discuss different health problems, for which carbohydrates are directly or indirectly responsible. With them, you will see a mountain of evidence showing that our thinking about carbohydrates, up until this time, has been terribly wrong.

Finally, I am going to give you my personal opinions about this important subject: Carbohydrates Can Kill!

Chapter 1: Nutrition And Health

NUTRITION 101

The human body needs vitamins, essential oils (fats it needs but does not produce), essential amino acids (amino acids it needs but does not produce), metals, and oxygen. Like a piece of machinery or a car, it also needs energy to keep it running. Food is the source of that energy.

A car runs on fuel and a machine runs on a power source such as electricity. It is all very straightforward, provided the fuel and the power meet the necessary quality requirements. However, if the fuel is somehow tainted, it can harm the engine, damage the car, and shorten the car's life. If the power that the machine uses is too strong or unsteady, the machine will break down and need replacement sooner. Likewise, foods provide energy to the body, but they may also cause diseases, and shorten life.

Why do some of us live healthier and longer than others do? Why do some of us suffer certain diseases while others do not? Sure, we can blame most of the diseases on our genes. We think that genes pass on the good and bad body characteristics from one generation to the next. However, do we really know why we find new genes in one generation, those not found in the earlier generations? This is called *mutation.* Do we really know the cause(s) for mutation or changes in genes? We must realize that we are what we eat! Overeating certain foods can be blamed for our poor health and premature death, perhaps by causing this mutation.

There are many kinds of foods. Society has given appealing, fancy names to many dishes, while it has presented others to us as "good for us" or "healthy." The only truly meaningful method of evaluating foods we eat is by considering the content of three major nutrients, or *macronutrients*. They are carbohydrate, protein, and fat. Our body needs each of them in a certain proportion, or more appropriately, a certain amount. Eating more or less than the correct amount of any nutrient can cause acute or chronic health problems. For example, many research findings have shown us that carbohydrate-rich diets are harmful.

I have prepared a brief tour through the subject of nutrition and the body. You will see how our body handles foods. At the end of the tour, you will understand how carbohydrate, protein, and fat affect our health.

In chemistry, we use the words element, molecule, and compound. An *element* in chemistry is a type of atom, distinguished by its atomic number. It cannot be further divided by any chemical technique. A *molecule* is a group of two or more atoms, and it is electrically stable. In the living body scientists use either molecule or *biomolecule*. A *compound* is a chemical combination of more than two elements or atoms.

Breaking down molecules of carbohydrates, proteins, and fats produces energy, which the body needs. A *calorie* is a measurement of energy. It is the amount of energy or heat needed to raise the temperature of a gram of water by

1°C (one degree Celsius) or the temperature of 0.00005 ounce of water by 1°F (one degree Fahrenheit). One calorie is such a very small unit that we typically use nutritional calories or kilocalories (1,000 calories) in describing the energy in foods. Under the ideal conditions, each gram of carbohydrates gives 4 (nutritional) calories or kilocalories. Each gram of proteins gives 4 (nutritional) calories or kilocalories. Each gram of fats gives 9 (nutritional) calories or kilocalories.

Carbohydrates

A *carbohydrate* (carbo-hydrate) is a molecule. It has carbon (C), hydrogen (H), and oxygen (O) atoms. *Carbo* is carbon. A *hydrate* is a bond between hydrogen and oxygen. Breaking down a carbohydrate molecule commonly produces carbon dioxide, water, and gives 4 (nutritional) calories or kilocalories.

Commonly, we group carbohydrates into simple carbohydrates and complex carbohydrates. *Simple carbohydrates* are the molecules with the smaller number of carbon atoms. *Complex carbohydrates* are the molecules with several units of simple carbohydrates. Breaking up a complex carbohydrate produces several simple carbohydrates. Examples of simple carbohydrates are sugar, flour, and their products. Those of complex carbohydrates are starch and fibers. The body can break up only a very few fibers in the digestive track for taking into the circulation. Fibers are good for building up the volume of our meals and the bulk of our bowel content. Generally, fibers do not affect our blood glucose (sugar) concentration.

Based on the structure of the molecules, we also group

carbohydrates into monosaccharide, disaccharide, or polysaccharide. *Saccharide* means "sugar," *mono* means "single," *di* means "double," and *poly* means "multiple."

Monosaccharide is a single sugar molecule; it is the smallest and simplest molecule for our digestive tract to take in. Glucose, fructose (fruit sugar), and galactose (brain sugar) are monosaccharides. The body can only take monosaccharides into its circulation, and use them for producing energy.

Disaccharide is a sugar molecule with two monosaccharides bonded together. Lactose (milk sugar) is an example of disaccharide; it contains glucose and galactose.

Polysaccharide is a large sugar molecule with more than two single sugar molecules bonded together. Starch and fibers are examples of polysaccharides. The body has to break disaccharides and polysaccharides up into monosaccharides, before it can take any into circulation through the digestive tract.

Proteins

A *protein* is a molecule of many amino acids, which are the basic units. Amino acid is a molecule of a carbon with links to an amino group (NH_2), a carboxylic acid group (COO), a hydrogen atom (H), and a side chain. An amino group includes a nitrogen atom (N) and two hydrogen atoms (H_2). Amino acid is the smallest molecule and the simplest form for absorption through the digestive tract. The body cannot produce some amino acids, which are essential to it. Hence, these are named *essential amino acids*.

Peptide is a molecule of two or more amino acid molecules, which are linked together. A *polypeptide* is a single linear (straight line) chain of amino acids. Protein is made up of one or more polypeptides, and includes more than 50 amino acids.

Proteins are the main supply of nitrogen to our body for tissue repair and replacement. Our body can break up proteins and convert them into carbohydrates for producing energy. One gram of proteins is supposed to give 4 kilocalories of energy. However, it takes a loss of energy to use proteins as an energy source. Each gram of proteins loses one kilocalorie in the process. This is the reason why replacing carbohydrates in diets with proteins, gram for gram, helps the user lose weight.

Fats

The term *fats*, when discussing health and nutrition, usually refers to *triglycerol* and *fatty acids*. These are the energy source for many tissues of our body. Excessive fats are stored as *triacylglycerols (triglycerides)*. Like amino acids, some fatty acids are essential to the body, because the body cannot produce them. These fatty acids are named *essential fatty acids*.

Some fats, at normal temperature, are solid, while oils (also fats) are liquid. Depending on the bonding between the carbon atoms, these are *saturated* or *unsaturated* fats. Saturated fats are mostly animal fats. Unsaturated fats are mostly vegetable fats.

The fat foods include *triglycerides* and *cholesterol*. Triglyceride is a triglycerol bonded with three fatty acids. However, not all the triglycerides and cholesterol in the blood circulation are from the fat foods. In fact, the more the carbohydrate foods we eat, the more the triglycerides and cholesterol our liver produces and releases into the circulation.

Most doctors and nutritionists blame the solid and saturated fats for clogging and sticking to the blood vessels, especially the arteries. This can cause arteriosclerosis and atherosclerosis, as well as blockage of the arteries. These are the reasons why we are advised to avoid as many saturated fats and cholesterol as possible.

DIGESTION & METABOLISM

Later in this book, I am going to show you research findings that detail ways, in which carbohydrate-rich diets are harmful. To prepare you for better understanding of the research findings, I am going to show you the digestive system. Then, I shall show you how foods work inside the body, both for good and for bad. Now, let me take you on a tour through the digestive system.

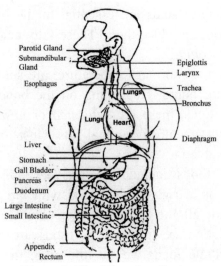

Figure 1. The Digestive System

First of all, teeth bite the food as soon as it passes into the mouth. The teeth also chew the food into small pieces, so that it can easily go into the esophagus. While the teeth are chewing, the tongue helps mix the food with saliva coming out of the salivary glands. The *salivary glands* include the parotid glands and the submandibular glands. The parotid glands are located in our both cheeks in front of the ears; and the submandibular glands are underneath the tongue extending to both rims of the mandible or lower jawbone.

The saliva contains water, mucin (mucus), and enzymes, which include alpha-amylase and lipase. Enzymes promote chemical reactions inside the body. *Alpha-amylase* breaks carbohydrates into disaccharide and monosaccharide. *Lipase* breaks triglycerides into fatty acids and *monoglyceride*.

As soon as the teeth finish chewing the food, the tongue moves backward to send the food into the esophagus. At the base of the tongue, there is a piece of cover, or epiglottis. The *epiglottis* falls down to cover the *glottis*, which is the opening of our respiratory (breathing) tract or *larynx*, when the tongue sends the food into the esophagus. This neatly coordinated action helps deter foods and fluids getting into our respiratory tract. Swallowing foods and fluids into the respiratory tract (*aspiration*) can be very dangerous and life-threatening because it interferes with breathing.

Now, let's go back to the subject of digestion. The *esophagus* is just a passage with peristalsis. *Peristalsis* involves contraction of the muscles of the esophagus, like tidal waves, to move foods into the stomach. With the mucus and water from the saliva, the food normally passes through the esophagus smoothly.

A dome-shaped sheet of muscles, the *diaphragm*,

separates the chest cavity and the abdominal cavity. It also separates the esophagus from the stomach. The diaphragm contracts in helping expand the lungs to breathe in air. At the same time, the contraction closes the passage between the esophagus and the stomach. The closure prevents the food and the fluid inside the stomach from going back into the esophagus.

The *stomach* has two types of cells: the parietal cells and the chief cells. The *parietal cells* produce *hydrochloric acid*, which is stomach acid. The *chief cells* produce *pepsinogen*, which becomes pepsin. *Pepsin* is an enzyme for breaking proteins into shorter and smaller peptides. The stomach is also the place for mixing the food further to make fats into *globules*.

After leaving the stomach, the food moves into the first part of the *small intestine*, which is the duodenum. An opening on the wall of the *duodenum* receives secretions from both the gallbladder and the pancreas. The rest of the small intestine is jejunum and ileum.

The *gallbladder* is a sac or reservoir for saving and concentrating the bile juice from the *liver*. This organ squeezes bile into the duodenum in helping digest fatty food. When someone has avoided eating fatty foods for a long time, his gallbladder has continued to concentrate the bile juice. When he later eats fats again, he could experience pain from a gallstone formed by this concentration.

The *pancreas* has both exocrine and endocrine glands. The *exocrine* glands mainly produce and release the enzymes, in helping digest the food inside the small intestine. The enzymes include *trypsin*, *chymotrypsin*, *pancreatic lipase*, and *pancreatic amylase*. In addition, the pancreas also discharges solutions, which contain lots of *carbonate* and *salt*. The *endocrine* glands produce *insulin, glucagon, somatostatin, gastrin*, and *pancreatic polypeptide*. In the process of digestion, the exocrine glands play a main role.

The food, bile salt, and the pancreatic enzymes leave the duodenum and continue to travel through the rest of the small intestine. During the passage, the bile salts and enzymes convert the food into the smallest and simplest form of nutrients for absorption by the cells of the intestinal wall. By the time the food gets to the end of the ileum and the beginning of the large intestine or colon, only fibers and water remain.

The colon continues to move the food remains toward the rectum while it absorbs the water. The fiber helps hold the water and keep the stool from hardening inside the rectum before moving out of the bowel.

The Digestion Of Different Nutrients

Now, let us look into how our body digests carbohydrates, proteins, and fats.

As mentioned earlier, carbohydrates include polysaccharides, such as starch; disaccharides such as sucrose (table sugar) and lactose (milk sugar); and monosaccharides, such as glucose and fructose (fruit sugar.) As the foods get inside the mouth, alpha-amylase in the saliva mixes with carbohydrates. The enzyme breaks some of them down into shorter polysaccharides, disaccharides, and monosaccharides. There is no chemical reaction for carbohydrates inside the esophagus and the stomach.

When polysaccharides, disaccharides, and

9

monosaccharide move into the duodenum, they meet with the enzymes from the pancreas such as *maltase, lactase,* and *pancreatic alpha amylase*. The enzymes break them into monosaccharide, which include glucose, galactose, and fructose. The cells of the small intestine take in the monosaccharides and transfer them into the blood circulation.

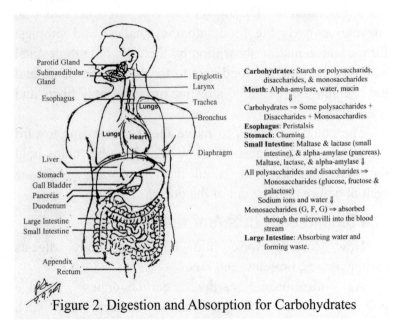

Carbohydrates: Starch or polysaccharids, disaccharides, & monosaccharides
Mouth: Alpha-amylase, water, mucin
⇩
Carbohydrates ⇒ Some polysaccharides + Disaccharides + Monosacchardies
Esophagus: Peristalsis
Stomach: Churning
Small Intestine: Maltase & lactase (small intestine), & alpha-amylase (pancreas).
Maltase, lactase, & alpha-amylase ⇩
All polysaccharides and disaccharides ⇒ Monosaccharides (glucose, fructose & galactose)
Sodium ions and water ⇩
Monosaccharides (G, F, G) ⇒ absorbed through the microvilli into the blood stream
Large Intestine: Absorbing water and forming waste.

Figure 2. Digestion and Absorption for Carbohydrates

As mentioned, no chemical reaction happens for proteins inside the mouth and esophagus. Inside the stomach, hydrochloric acid breaks protein or multiple polypeptides into long, single polypeptides; pepsin breaks the long, single polypeptides into short polypeptides.

The small intestine produces enterokinase, which helps convert the pancreatic trypsinogen into trypsin. In turn, trypsin helps activate other protein digesting enzymes from the pancreas. The pancreatic protein digesting enzymes break all the long and short polypeptides into amino acids. The cells of the small intestine take in amino acids and transferred them into the blood stream.

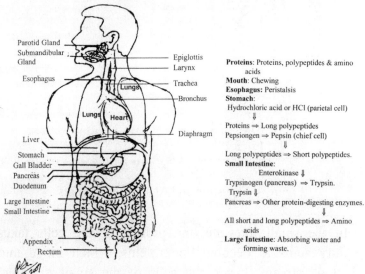

Figure 3. Digestion and Absorption for Proteins

Little changes for fat foods inside the mouth and esophagus, although there is an enzyme, lipase, for fats in the saliva. After getting into the stomach, the mass of fats is blended into large fat globules before it is moved into the duodenum. The duodenum receives bile salt in the bile from the gallbladder and lipase from the pancreas. Bile salt acts like soap to break large fat globules into small fat globules. Lipase breaks small fat globules into monoglyceride and free fatty acids. Triglycerol is a molecule of three carbons bonded together with three hydrates (OH). Monoglyceride is a triglycerol with one fatty acid.

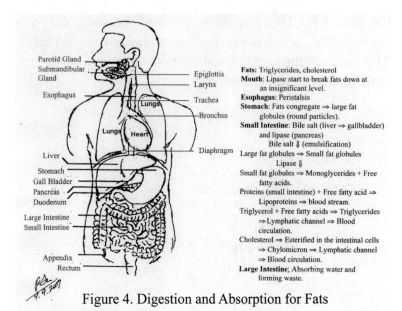

Parotid Gland
Submandibular Gland
Esophagus
Lungs
Lungs
Heart
Liver
Stomach
Gall Bladder
Pancreas
Duodenum
Large Intestine
Small Intestine
Appendix
Rectum

Epiglottis
Larynx
Trachea
Bronchus
Diaphragm

Fats: Triglycerides, cholesterol
Mouth: Lipase start to break fats down at an insignificant level.
Esophagus: Peristalsis
Stomach: Fats congregate ⇒ large fat globules (round particles).
Small Intestine: Bile salt (liver ⇒ gallbladder) and lipase (pancreas)
Bile salt ⇓ (emulsification)
Large fat globules ⇒ Small fat globules
Lipase ⇓
Small fat globules ⇒ Monoglycerides + Free fatty acids.
Proteins (small intestine) + Free fatty acid ⇒ Lipoproteins ⇒ blood stream.
Triglycerol + Free fatty acids ⇒ Triglycerides ⇒Lymphatic channel ⇒ Blood circulation.
Cholesterol ⇒ Esterified in the intestinal cells ⇒ Chylomicron ⇒ Lymphatic channel ⇒ Blood circulation.
Large Intestine; Absorbing water and forming waste.

Figure 4. Digestion and Absorption for Fats

In the small intestine, the protein of the cells takes up fatty acids to become lipoproteins, which move into the blood stream. Inside the cells of the small intestine, monoglyceride and fatty acids bond themselves to become triglycerides. Triglycerides go through the lymphatic channel into the blood circulation.

The Metabolism Of Different Nutrients

After all the *macronutrients* (carbohydrate, protein, and fat) enter the blood circulation, they go to the cells, tissues, or organs. Please see Figure 5 below. You do not have to understand or remember everything in the figure. You just follow the arrows and connect the important names; you will understand the relationships among the names.

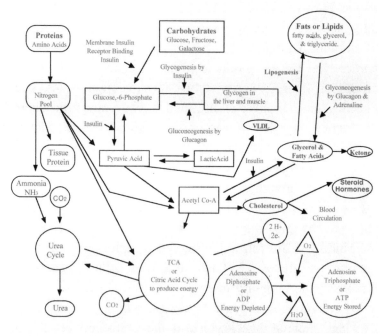

Figure 5. Metabolism of the Nutrients

Amino acids reach the cells and become protein for tissue replacement and repair. This is the process of *anabolism*. When the body does not have enough supply of glucose for energy production, the body breaks protein into amino acids for producing energy. This process is *catabolism*. Amino acids may enter the TCA cycle for energy production; and may be recycled to form new amino acids.

Fatty acids and triglycerides, reach the fat cells (adipocytes) for storage. Cholesterol is usually found in the cellular membrane of all animals and some plants. When the body runs out of energy supply from carbohydrates or sugars, the body breaks fats down to glycerol and free fatty acids, which enter the TCA cycle for energy production.

13

However, it is impossible to convert glycerol and fatty acids into glucose, except some fatty acids with an even number of carbons. Therefore, *hydroxybutyrates*, or a kind of *ketone bodies*, are the by-products in the breakdown of fats for producing energy. Fortunately, the skeletal muscles, heart muscles, and brain can use hydroxybutyrates for energy production. Another ketone body, *acetoacetate*, is produced in the situation of uncontrolled hyperglycemia and diabetic acidosis. Acetoacetate cannot be used by the body and has to be removed through the kidneys.

Glucose makes up the largest portion of the blood sugars absorbed by the small intestine. It reaches the cells, with the binding of insulin to the insulin receptors of the cellular membrane; glucose is taken into the cells. (Pleases see Insulin in the Glossary of Important Terms.) Insulin further helps convert glucose into glycogen, which is stored in the liver and skeletal muscle. Insulin also helps convert glucose into pyruvate and acetyl co-enzyme A (Acetyl CoA). Acetyl CoA enters the TCA cycle for energy production. With the help from insulin, through the formation of Acetyl CoA, glucose is used to produce fatty acids, glycerols, triglycerol, and cholesterol. Fatty acids, joined by triglycerols, become triglycerides. With protein, triglycerides and cholesterols form *VLDL* (very low-density lipoprotein.) When the amount of glucose is more than the need for energy production, the levels of fatty acids, triglycerides, cholesterols, as well as VLDL increase. As a result, the body gains fats.

GLOSSARY OF IMPORTANT TERMS

Later in this book, as I show you the research findings about carbohydrate-rich diets, you will want to have a better understanding of some of the terms I use. Below is a list of terms for your reference.

Alpha Cell of the Pancreas is one of the five types of the endocrine cells of the pancreas. Endocrine cells produce hormones. The mass of alpha cells is 15-20% of the total islet of Langerhans cells of the pancreas. Alpha cells produce glucagon. Glucagon increases blood glucose by converting glycogen and other nutrients into glucose. For more information see **Pancreas.**

Amino Acid is the smallest unit of the protein family. Its molecule contains both amine (NH_2) and carboxyl (COOH) groups. Amino acids are used to build proteins. Amino acids can be used in producing energy. Some amino acids can be converted into glucose. Essential amino acids, which our body cannot produce, are supplied from foods. See the figure of metabolism. For more information, see **Protein.**

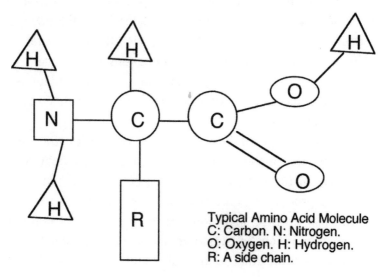

Typical Amino Acid Molecule
C: Carbon. N: Nitrogen.
O: Oxygen. H: Hydrogen.
R: A side chain.

Figure 6. Typical Amino Acid Molecule

Amyloid-beta (ß-amyloid) is a fibrous protein, after changing the linkage between the protein molecules by glycation. The fibrous protein aggregates to become insoluble (not dissolved in water) and massive plaque. The plaque formation around the brain cells is the cause of Alzheimer's disease.

Amyloidosis is a variety of conditions, in which amyloid proteins are abnormally deposited in the tissues and organs, generally or locally. Amyloidosis is found in diseases such as Alzheimer's disease, Parkinson's disease, ALS (Lou Gehrig's disease), multiple sclerosis, and so on.

Arteriosclerosis is a disease, in which the artery becomes stiff and hard. First, the inside layer of the artery (endothelium) gets inflamed, usually as a result of high blood glucose (*hyperglycemia*). Inflammation triggers the deposit of cholesterol from the circulating LDL-c (low-density lipoprotein cholesterol), especially the VLDL-c (very low-density lipoprotein cholesterol). Inflammation also promotes the accumulation of macrophage, a kind of white blood cells. When the endothelium of an artery is deposited with cholesterol plaques (atheromatous plaques), the artery is hardened. **Atherosclerosis** is an advanced type of arteriosclerosis.

Beta (ß) Cell of the Pancreas is one of the five endocrine cells of the pancreas. The mass of the beta cells is 65-80% of the islet of Langerhans cells of the pancreas. Beta cells produce insulin. Insulin lowers blood glucose by helping use glucose in producing energy. It helps convert glucose into glycogen stored in the liver and muscles. It also helps convert glucose into fatty acids and glycerol (fat) stored in the fat tissue.

Body Mass Index (BMI)) is the product by dividing one's body weight by his total body surface. The square of a person's height equals his total body surface. Although we often mention BMI by its number, its unit is either Kg (kilograms)/ M^2 (square Meter) or pounds X 703/ $inch^2$ (square inch.)

Carbohydrate is a molecule of carbon, hydrogen, and oxygen. It is sugar or saccharide. Carbohydrates are one of three major nutrients (macronutrients) for our body. The others are proteins and fats. Carbohydrates include monosaccharide, disaccharide, and polysaccharide. Monosaccharide is the smallest unit of the family. Disaccharide is a molecule with two monosaccharide bonded together. Polysaccharide is a molecule with more than two monosaccharides bonded in the molecule. Starch, glycogen, chitin (of animals), and cellulose (of plants) are polysaccharides.

Carbohydrate

C: carbon. H: hydrogen. O:oxygen. The number of carbon, hydrogen, and oxygen in each molecule varies.

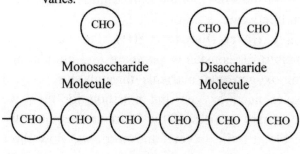

Monosaccharide
Molecule

Disaccharide
Molecule

Polysaccharide
Molecule

Figure 7. Typical Molecules of Carbohydrates

Cholesterol is one kind of lipid, and a major component of the cellular membrane of all animals. It is also used to produce steroid hormones including sexual hormones. A trace of cholesterol is also found in the plants and fungi. Our body produces most of the cholesterol from foods, especially from glucose or carbohydrates. Cholesterol is found very much in the liver, spinal cord, and brain. It is insoluble in the blood. It is only transported by the water-soluble lipoproteins in the blood circulation, especially in the LDL (low-density lipoprotein) and VLDL (very low-density lipoprotein). The amount of cholesterol inside LDLs, especially VLDL, is linked to arteriosclerosis or atherosclerosis and coronary heart disease.

Collagen is the most important protein of the connective tissues in animals, particularly mammals. The long and tough fibrous proteins are called collagen fibers, which are abundant in the tissues between cells to support the structure of the tissues. Collagen fibers are the major component of cartilage, ligaments, fascia (the cover of the muscle), tendons, bones, and skin. Collagen fibers and keratin are responsible for the stretch of the skin. Aging or degeneration of collagen fibers causes skin wrinkles.

Coronary Heart Disease (CHD): The heart muscles are responsible for constantly pumping and circulating the blood with oxygen and nutrients through our body. The heart muscles need oxygen and nutrition for their work, too. The coronary arteries of the heart are responsible for letting the blood (with oxygen and nutrients) flow through the heart muscles. The coronary heart disease is a result of accumulation of atheromatous plaques, which are built up with cholesterol. The plaques narrow the passage for the blood flow inside the arteries. When the blood cannot flow through the heart muscles (ischemia); the heart muscles cannot work normally and will die. CHD is a leading cause of death in the US.

Diabetes Insipidus (DI) is caused by the deficiency in the anti-diuretic hormone (ADH.) The posterior lobe of the pituitary gland secretes ADH; the pituitary gland is located inside our brain. The patient with this disease cannot concentrate his urine and continues to lose his body water into the urine. DI can cause dehydration and hypotension. DI is mentioned here is to distinguish it from diabetes mellitus (DM.)

Diabetes Mellitus (DM) is diagnosed when the level of blood glucose is above a set of limits used during blood testing. Such a diagnosis indicates that the pancreas is unable to bring down the level of blood glucose as expected. There are three tests used for making diagnosis with diabetes mellitus.

(1) Fasting Plasma Glucose (FPS) Test: The level of the blood glucose in the morning after fasting at least eight hours and before having breakfast. Note: The plasma concentration in the following charts, mg%, means the number of milligrams of glucose in 100 milliliters of blood (or plasma).

Plasma Glucose	Diagnosis
99 mg% or below	Normal
100-125 mg%	Pre-diabetic
126 mg% and over	Diabetic

Figure 8. Diagnostic Criteria for Diabetes Mellitus (1)

(2) Glucose Tolerance Test (GTT): The level of blood glucose at two hours after drinking a solution of 75 grams of glucose. Diagnosis is made after having abnormal results from two tests done on two different days.

2-hour Plasma Glucose	Diagnosis
139 mg%	Normal
140 mg% -199 mg%	Pre-diabetic
200 mg% and over	Diabetic

Figure 9. Diagnostic Criteria for Diabetes Mellitus (2)

(3) Gestational Diabetes Test: It is a condition with abnormal GTT results during pregnancy, with additional readings at the end of the first and third hour during the test. Diagnosis is made after having abnormal results from two tests done on two different days.

Time When Blood Samples Taken	Gestation Diabetes
Fasting	99 mg% or above
At the end of first hour	180 mg% or above
At the end of second hour	155 mg% or above
At the end of third hour	140 mg% or above

Figure 10. Diagnostic Criteria for Diabetes Mellitus (3)

There are three main types of diabetes mellitus, including type 1, type 2, and gestational diabetes.

Type	Characteristics
Type 1	Acute onset in child, teenager, or young adult. The patient is insulin dependent.
Type 2	Gradual onset in adult. Initially, the patient is non-insulin dependent, but may become insulin dependent.
Gestational	Onset in the late stage of pregnancy. After the end of pregnancy, the patient may recover or become type 2 DM.

Figure 11. Types of Diabetes Mellitus

Enzyme is a molecule, mostly protein and naturally produced in living organisms. It enables chemical reactions inside the organisms.

Erectile Dysfunction (ED) is a condition, in which the male penis or female clitoris fails to become erect in response to stimulation. There are different causes for erectile dysfunction. However, erectile dysfunction may likely be a symptom of a host of cardiovascular diseases.

Fat(s) and Lipids include fats, cholesterol, oils, and waxes. Often, we refer fats for all lipids except waxes. Strictly, fats are glycerides. Glycerides have a basic structure with bonding between a glycerol (glycerin) and one, two or three fatty acid(s). Depending on the number of fatty acids, which are bonded to the glycerol, the names of the glyceride molecules are monoglyceride (one fatty acid), diglyceride (two fatty acids), and triglyceride (three fatty acids), respectively. Depending on the saturation of the bonding between the carbons of the fatty acids, fats are classified into saturated fats (all single bonds) and unsaturated fats (at least one double bond). Generally, animal fats are saturated; and the plant fats or oils are unsaturated. However, there are some plant saturated fats too.

Fatty acid is an acid molecule of four carbons or more, plus oxygen and hydrogen. Our body cannot produce essential fatty acids, which must come from foods. The body produces some fatty acids from glucose. Depending on the saturation of the bonding between the carbons, fatty acids are grouped as saturated and unsaturated.

Fructose is the sugar of fruits, and is very plentiful in syrups, honey, and table sugar. Sucrose is a disaccharide molecule with a glucose molecule and a fructose molecule bonded together.

Galactose is a monosaccharide, the simplest unit of sugars, mainly found in the brain, and not as sweet as glucose. It bonds itself to glucose to become lactose, a disaccharide.

Glucagon: Please see **Alpha Cell of the Pancreas** and **Pancreas**.

Glucose is a simplest unit of sugars or monosaccharide. It is the major component of blood sugar. It bonds itself with galactose to become lactose, and with fructose to become sucrose. For more information, see **Carbohydrate**.

Glycation is a simple bonding reaction between sugar and fat, protein, or both. There are enzymatic glycation and non-enzymatic glycation.

Glycation

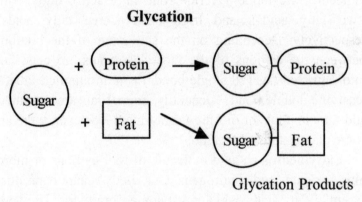

Glycation Products

Figure 12. Glycation

Enzymatic glycation is a bonding process, which requires the presence of enzyme(s). It is also called glycosylation. A typical glycosylation is the bonding between the protein part (globin) of hemoglobin, and glucose in the blood circulation. The percentage of globin of the hemoglobin, which is bonded with glucose, out of the total amount of globin in the circulation is expressed as hemoglobin A_{1c} or HbA_{1c}. For more information, please see **Hemoglobin A_{1c}**.

Glycosylation

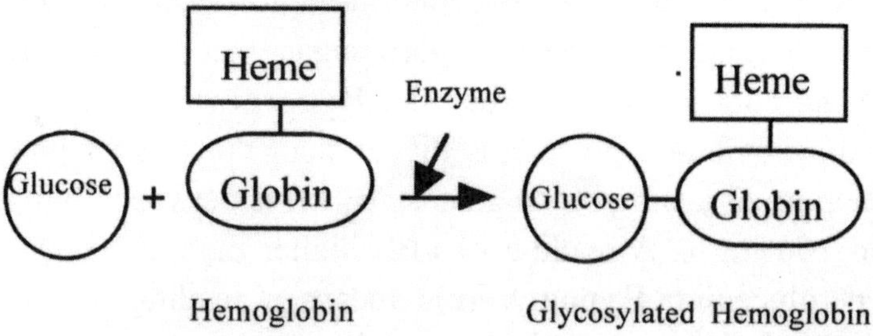

Figure 13. Glycosylation

Non-enzymatic glycation is a bonding between sugar, and protein or fat or both, without the need of an enzyme. This process happens in both outside and inside our body. Heat is supposed to help the process. The product of glycation is called a glycation product. The sugar part of the glycation product is glycan. Glycation product is toxic to our tissues. Glycation product can be a seed for further glycation like that in making a snowball. The glycation product can continue to grow bigger. The bigger product is named advanced glycation endproducts (AGEs).

Glycemia is the concentration of sugar of the blood. **Glycemic index** is the strength of a carbohydrate food to increase the concentration of sugar in the blood, after the body digests and absorbs the food. The glycemic index of a fixed amount of pure sugar or glucose is expressed as 1.00 (100%). As compared to eating one gram of pure sugar or glucose, the rise of the level of blood glucose is a food's glycemic index, after eating one gram of the carbohydrate food. The higher the glycemic index of a carbohydrate food is, the higher the level of blood sugar will be, after digestion of the food. **Glycemic load** is the product of a food's glycemic index and the consumed amount of the food. For example, the glycemic index for sugar is 1.00. If a person had 100 grams of pure sugar, the glycemic load that he consumed would be 100. **Hyperglycemia** means a reading of blood glucose is higher than the normal limits. The normal fasting blood glucose (FBS) is between 70 mg% and 100 mg%. A reading of FBS higher than 100 mg% is hyperglycemia. **Hypoglycemia** means a reading of blood glucose is lower than the normal limits. A reading of FBS lower than 70 mg% is hypoglycemia. For more information, please see **Diabetes Mellitus**.

    Glyceride is a molecule of glycerol and fatty acid(s). Depending on the number of fatty acids bonded to the glycerol, glycerides include monoglyceride, diglyceride, and triglyceride. For more information, please see **Fat(s)**.

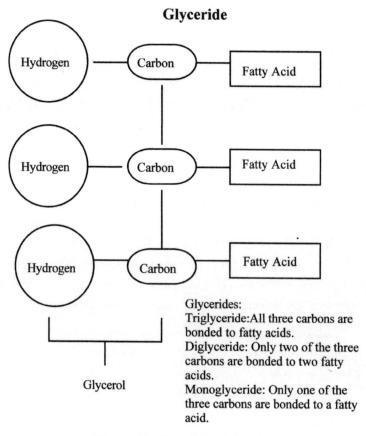

Figure 14. Glyceride Molecule

Glycerol or glycerine is the backbone of the molecule of glyceride. Chemically, it includes three carbon atoms (C), three hydrogen atoms (H), and three hydroxyl groups (OH.) The hydroxyl groups make glycerol soluble in water. They are replaced by fatty acid(s) to form glycerides. Other than coming from food, the body produces glycerol and fatty acid from glucose. This is the reason why eating a high-carbohydrate diet will place a high amount of triglycerides and fatty acids in the blood circulation.

Hemoglobin A$_{1c}$ or HbA$_{1c}$ is the percentage of glycosylated hemoglobin of the total circulating hemoglobin.

HbA$_{1c}$ = Glycosylated Hemoglobin ÷ Total Circulating Hemoglobin X 100%

Or, HbA$_{1c}$ is the percentage of globin of the hemoglobin, which is bonded with glucose (a process of enzymatic glycation), as compared to the total amount of globin in the circulation. The higher the concentration of the blood glucose is, the higher the percentage of the hemoglobin A$_{1c}$ or HbA$_{1c}$ will be. This is the reason that we use the reading of HbA$_{1c}$ to estimate the level of blood glucose and the severity of diabetes mellitus (DM). For more information about enzymatic glycation please see entries under **Glycation**.

The conversion between Hemoglobin A$_{1c}$ and blood glucose level is

1 mg% (blood glucose) = 0.055 mml/L= 0.02857% (Hemoglobin A$_{1c}$)

Blood Glucose in mg%	Blood Glucose in mmol/L	Hemoglobin $_{A1c}$ in %
65	3.5	4
100	5.5	5
135	7.5	6
170	9.5	7
205	11.5	8
240	13.5	9
275	15.5	10
310	17.5	11
345	19.5	12

Figure 15. Hemoglobin A$_{1c}$ and Blood Glucose Levels

HDLs (High Density Lipoproteins) appear in various sizes. They have much more protein (40-55%) and less cholesterol and triglycerides (50-55%) than LDLs, VLDL, or chylomicron does. For more information please see **Plasma Lipoproteins**.

Hypertension meant high blood pressure. There are two readings of blood pressure; the top is the systolic blood pressure; the bottom is the diastolic blood pressure. The systolic blood pressure is the pressure when the heart squeezes the blood out of its left lower chamber (ventricle) into the aorta. The diastolic blood pressure is the pressure when the heart is relaxed after squeezing the blood out of its left ventricle, and closes its aortic valves. The average blood pressure readings for a normal, healthy person are 120/70 mmHg (the millimeter of Mercury.) The upper limits of blood pressure for a normal, healthy person are 140/90 mmHg. Hypertension is diagnosed when a reading of either the systolic or diastolic pressure, or both, are higher than their normal limits.

Impotence is erectile dysfunction. There are different causes for impotence or ED. It is a diabetic complication. When impotence or ED happens before discovering any disease, it is important to have a screening for diabetes mellitus and the cardiovascular system. For more formation, see **Erectile Dysfunction**.

Insulin is a hormone produced by the beta cells of the Islets of Langerhans. Insulin is required in several stages of disposing glucose; therefore, it is important for keeping the blood glucose level within the normal range.

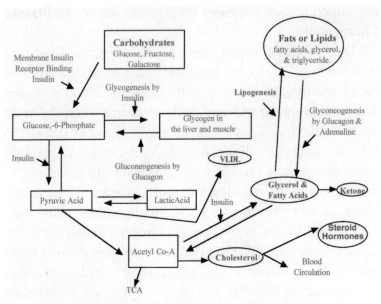

Figure 16. Insulin and Glucose Metabolism

Insulin reaches the cells, and couples itself with insulin receptors (figure below) of the cellular membrane. Insulin helps cells take in glucose in the circulation. It helps store glucose in the liver and muscles as glycogen. It helps break down glucose-6-phosphate to acetyl co-enzyme A (acetyl CoA), through pyruvic acid for generating energy in the TCA cycle, and produces fats (fatty acids and glycerols) and cholesterol.

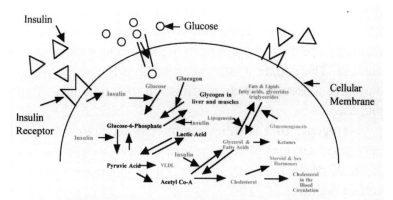

After insulin is coupled with the insulin receptor of the cellular membrane, insulin helps the cell take in glucose; store glucose in the liver and muscles as glycogen; process glucose-6-phosphate to acetyl Co-enzyme A for generating energy in the TCA cycle; and produce fats (fatty acids and glycerols.)

Figure 17. Insulin and Cellular Utilization of Glucose

Insulin Resistance is a complication of a diabetic patient, when he continues to have high blood glucose, despite that he has a high concentration of insulin in his blood circulation. This situation is thought that the insulin receptor of the cellular membrane becomes insensitive to insulin. For more information, see **Insulin** and **Pancreas**.

Ketoacidosis or Ketosis happens when the body does not produce insulin or cannot use insulin in the circulation, its liver breaks down proteins and fats for producing energy and ketone bodies. Ketone bodies include acetone, acetoacetate, and beta-hydroxybutyrate. As the body produces more ketone bodies than it can use, it will be in the state of ketoacidosis. Ketoacidosis is observed in the crisis of type 1 diabetes mellitus, in the advanced stage of type 2 diabetes, as well as in starvation. In ketoacidosis, the concentration of free fatty acids in the circulation rises. However, the patient with diabetic acidosis has high concentration of blood glucose, while the one with ketoacidosis from starvation or a low carbohydrate diet does not have high concentration of blood glucose. The people in starvation or on a low carbohydrate diet have a normal function of insulin to handle their blood glucose.

Ketogenic Diet is a diet without carbohydrates or with a very small amount of carbohydrates. It can trigger the liver to break down fats and proteins for producing energy and ketone bodies. .

Ketone Bodies include acetone, acetoacetate, and beta-hydroxybutyrate. The heart muscles, skeletal muscles, and brain can use beta-hydroxybutyrate as their energy source. However, the body cannot use acetone and acetoacetate, which are removed from the body in the urine.

Lactose is milk sugar and a disaccharide (a molecule with two single sugar molecules). The two single sugar molecules are glucose and galactose.

LDLs (Low Density Lipoproteins) occur in various sizes and have more protein than VLDL and chylomicron (15-25%) do. LDLS also have slightly less cholesterol and triglycerides (75-85%) than VLDL do. An abnormal increase in the amount of LDLs especially the VLDL increases the risk of coronary heart disease, arteriosclerosis, and atherosclerosis. For more information, see **Plasma Lipoproteins**.

Molecule is a substance that is electrically stable and made of two or more atoms.

Pancreas is an important organ involved in digestion and metabolism. There are endocrine and exocrine glands in the pancreas. The exocrine glands produce enzymes, which are sent into the duodenum for digesting carbohydrates, proteins, and fats. There are four major endocrine glands in the clusters of cells called *Islets of Langerhans*. The endocrine glands produce hormones and send them into the blood circulation. Based the types of hormones, there are five types of cells. They are (the number in the parentheses are the percentage of the total Langerhans cells) alpha cells (15-20%), beta cells (65-80%), delta cells (3-10%), PP cells (3-5%), and epsilon cells (less than 1%.)

Peptide is a molecule, which bonds two or more amino acids together. Depending on the length of the chains, there are short and long peptides or polypeptides. For more information see **Protein**.

Plasma Lipoproteins mainly transport fats in the circulation. The fluid component of the blood is plasma. Lipoprotein is a biochemical, in which lipids (lipid or fat) and protein are bounded together. Lipoproteins include toxins, enzymes, adhesins (substances to cause adhesion or sticking between cells), antigens (substances to trigger production of antibodies for immune reactions), cellular structure proteins, and transporters. Based on their densities, plasma lipoproteins include HDLs (high density lipoproteins), LDLs (low density lipoproteins), IDLs (intermediate density lipoproteins). VLDL (very low density lipoproteins), and Chylomicrons.

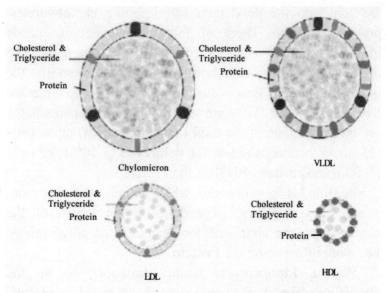

Figure 18. Types of Plasma Lipoproteins

As in the figure, protein forms the outside layer of the lipoprotein molecule; cholesterol and triglycerides fill the inside of the molecule. Chylomicron molecule has less protein (1.5-2.5% of the molecule) and more cholesterol and triglycerides (97-99% of the molecule) and is the least in density among all the plasma lipoproteins. VLDL has slightly more protein (5-10%) and slightly less cholesterol and triglycerides (90-95%) than chylomicron; and is slightly greater in density than chylomicron. LDLs are in various sizes. LDLs have more protein (15-25%) than VLDL and chylomicron; and have slightly less cholesterol and triglycerides (75-85%) than VLDL. HDLs in various sizes, and have much more protein (40-55%) and less cholesterol and triglycerides (50-55%). (From Textbook of Biochemistry with clinical correlations by Thomas M. Devlin.) The increase of LDLs, especially the VLDL, increases the risk of coronary heart disease, arteriosclerosis, and atherosclerosis.

Starch is a polysaccharide with both straight and branched chains of multiple glucose molecules (or polymers). For more information, see **Carbohydrate.**

Stroke is the damage(s) of the brain as a result of lack of blood circulation to it. Either a blockage or rupture of the blood vessel(s) to the brain can cause stroke.

Sucrose is table sugar and a disaccharide molecule with a bond between a glucose molecule and a fructose molecule. For more information see **Carbohydrate.**

TCA Cycle (Tricarboxylic Acid Cycle) is an energy-producing cycle or process of the body. To start the energy-producing process, it requires a small amount of oxaloacetate from the breakdown of glucose to react with acetyl co-enzyme A (acetyl CoA). Therefore, a diet with little or no carbohydrates cannot run the TCA cycle, and ends in the state of ketosis.

Triglyceride is a molecule with bonds among a glycerol and three fatty acids. It is the fats in the diet and blood circulation. However, our body makes triglycerides from glucose. This is why eating more carbohydrate foods raises blood glucose (not necessarily diagnosed with diabetes mellitus yet) and increases of triglycerides in his plasma (blood). For more information see **Fats** and **Glycerides.**

VLDL (Very Low Density Lipoprotein) is one of the lipoprotein family members. In comparing to chylomicron, it has slightly more protein (5-10%); and has slightly less cholesterol and triglycerides (90-95%); and is slightly greater in density. An increase in VLDL is most responsible for the higher risks of arteriosclerosis and atherosclerosis, and coronary heart disease. For more information see **Plasma Lipoproteins**.

Chapter 2: Overweight, Then Obese

THE POPULATION IS GETTING HEAVIER

I completed my residency program at the Medical College of Virginia at the end of June 1974, and began private practice in anesthesiology and pain medicine (pain management). I retired from practicing anesthesiology in late August 1997, but continued to practice exclusively in pain medicine with acupuncture at my office.

Anesthesiologists do not only have to be familiar with their important role in all surgical procedures. They must also be knowledgeable of all medical disorders and their impacts, during the course of surgery and the recovery from it.

Just set aside for now those notable medical disorders, which are critical to the anesthetic management, involving the heart, the lungs, the liver, the kidneys, and the brain. Securing an airway for the patient is the number one concern for the anesthesiologist throughout the period, between the time when the patient loses his consciousness with anesthetics, and the time when the patient regains his consciousness from anesthesia. Blocking the airway for slightly more than a couple of minutes can cause grave consequences, even death.

Other than a solid foreign body or fluids in the path of the airway or anatomical malformation, obesity is one of the most significant factors that impact the success of maintaining an unobstructed airway. An obese patient usually comes with a full and short neck. Too much fat around the neck, the oversized soft tissues combined with neck muscles, make it very difficult to stretch the neck up, and keep the airway unobstructed. In addition, an obese patient may have an oversized tongue. When he is sleeping on his back with or without anesthetics; his tongue falls onto the back of the mouth easily and readily blocks the airway.

Over the years of my practice in giving anesthetics for surgeries, I had observed changes in the American population. Among the most noticeable changes were the increases in both life expectancy and average body weight. When I started my anesthesiology residency training in 1972, the age of my "old patients" (geriatric) was most commonly between 60 and 70 years, and much rarely over 70 years. Since entering private practice, the age of my geriatric patients rose steadily. More and more, I was taking care of patients in the operating room who were 80 years old and up. By the time I retired from administering anesthetics in the operating room, I had had five patients who were over 100 years old.

I am glad to witness the continuing expansion of life expectancy. However, like my colleagues, I have seen a growing number of cases with problems of the heart, the lungs, the kidneys, the liver, and the brain. This phenomenon may reflect a natural course of life, in which vital organs increasingly become less efficient as part of the aging process. However, this phenomenon also shows two other significant facts.

- Continuing advancement in medical technology is saving more lives in emergency situations than it could before. As a result, more people are living longer.

- Medical technology, despite its continuing advancement, has concentrated on symptomatic treatment of inefficient organs. It has done very little in helping people prevent medical problems from building up in the first place. As a result, more people live longer with illnesses. At the same time, they are likely incapacitated physically, mentally, or both.

Being incapacitated seriously alters quality of life, so longevity under these circumstances would not be the preferred choice for most people. Many people who are incapacitated cannot perform everyday tasks without assistance from others. They usually wish that they did not need the assistance, that they could regain their former health. Often, they are living in limbo. From a socioeconomic standpoint, we are spending more and more money on healthcare each year. We have no hope of slowing down the continuing increase of health care costs based on current trends.

Many years ago, obesity was much more commonly seen among people who were wealthy. In some cases, obesity was seen as a symbol of their success and achievement. Until the first half of the 20th century, slight obesity had been thought a part of the natural aging process. It was more commonly seen among the older people of the population. At that time, young people who were overweight were few, and often more self-conscious about their appearance.

Thanks to advances in agricultural technology since the second half of the 20th century, foods have become more plentiful and readily available in the United States and other countries. They are less expensive, requiring a smaller portion of the average individual income. In the US, the cost decrease is largely due to government subsidy of agriculture.

As the population gets heavier, the people who are overweight or obese feel more comfortable in public. More people around them are also heavier. They have lost the peer pressure to maintain a lower weight, as well as the urgency to change their eating habits and lifestyle.

No one can put a brake on runaway health care costs. As a result, many people are suffering growing pressure in covering their health care expenses. Every year their health insurance premiums increase, even if they are young and healthy. But are they really as healthy as they think they are? Many people, who believe that they are healthy, continue to live an unmonitored lifestyle. They eat whatever they want and do whatever they wish. Suddenly, they find they have one or more serious health problems; the health care cost can be financially catastrophic to them. Their health problems are mostly related to their weight gain.

HOW DO WE DEFINE "OVERWEIGHT" AND "OBESE?"

During my years in medical school, I was trained to observe patients and to diagnose their problems, even before I began to talk to them and examine them. As a physician now, I cannot help but continue my habit of constantly observing people.

Sometimes, when I would sit on a bench outside a store,

waiting for my wife, I would watch people passing by. Some of them were slightly heavy, with a "pop-out" in their lower abdomen. Some of them were much heavier, with a large "tummy" from their breasts and down to their groins. Worst of all, a few had so much extra fat that it was hanging below their groins. I remember thinking that the extra weight must get in the way of their daily activities. I had a hard time understanding how those people could just put up with it every day, without thinking about getting rid of it.

Based on the National Health and Nutrition Examination Survey (NHANES), the percentage of the US adult population (between 20 and 70 years old), who were either overweight or obese, was 47% during the years between 1976 and 1980; 56% for the period between 1988 and 1994; and 65% during the period between 1999 and 2002. In the same survey, the percentage of the US population who were obese was 15% during the years between 1976 and 1980; 23% for the period between 1988 and 1994; and 31% during the period between 1999 and 2002. (Source: National Center for Health Statistics, US Department of Health and Human Resource.)

A very disturbing study has found that the epidemic of obesity has spread into the younger population as well. Based on the NHANES, the percentage of US child population (between 4 and 11 years of age) who were either overweight or obese was 4% during the years between 1971 and 1974, 7% for the period between 1976 and 1980; 11% during the period between 1988 and 1994, and 16% for the period between 1999 and 2002. In the same survey, the percentage of the US adolescent population (between 12 and 19 years of age, excluding the pregnant women), who were either overweight or obese was 6% during the years between 1971 and 1974; 5% for the period between 1976 and 1980, 11% during the period between 1988 and 1994, and 16% for the period between 1999 and 2002. (Source: National Center for Health Statistics, US Department of Health and Human Resource.)

Obesity in childhood is closely related to many diseases in adulthood. The serious ailments include diabetes mellitus (DM), cardiovascular diseases, arthritic diseases, cerebral vascular attacks (strokes), cancers, ovarian diseases, erectile dysfunction, and so on. (See Chapter 6)

Lately I have found that at least seven or eight out of every ten people are either overweight or obese. I have also observed a rapid increase in the number of children and young adults who are either overweight or obese. The numbers continue to rise at a shocking rate. With the knowledge of the health risks with overweight and obesity, this alarming fact is not one that should be ignored.

So, how do we define overweight and obesity? There are often disputes over the cutoff points when people move from one weight group into another. At this time, we still use Body Mass Index (BMI) to identify one's weight. From this index we determine what is below normal, normal, above normal (overweight) or excessively above normal (obese).

Body Mass Index (BMI) is the product of dividing one's body weight (kilograms) by the square (centimeters) of his height (his total body surface).

Body Mass Index

BMI	18	19	20	21	22	23	24	25	26	27	28	29	30	31	32	33	34	35
Height in Inch								Body Weight in Pounds										
58	86.1	90.9	95.7	100.5	105.3	110.1	114.8	119.6	124.4	129.2	134.0	138.8	143.6	148.3	153.1	157.9	162.7	167.5
59	89.1	94.1	99.0	104.0	108.9	113.9	118.8	123.8	128.7	133.7	138.6	143.6	148.5	153.5	158.5	163.4	168.4	173.3
60	92.2	97.3	102.4	109.1	114.1	119.0	124.0	128.9	133.9	138.8	143.8	148.7	153.7	158.6	163.6	168.5	173.5	178.4
61	95.3	100.6	105.9	111.2	116.4	121.7	127.0	132.3	137.6	142.9	148.2	153.5	158.8	164.1	169.4	174.7	180.0	185.3
62	98.4	103.9	109.4	114.8	120.3	125.8	131.2	136.7	142.2	147.6	153.1	158.6	164.0	169.5	175.0	180.4	185.9	191.4
63	101.6	107.3	112.9	118.6	124.2	129.9	135.5	141.1	146.8	152.4	158.1	163.7	169.4	175.0	180.7	186.3	192.0	197.6
64	104.9	110.7	116.5	122.4	128.2	134.0	139.8	145.7	151.5	157.3	163.1	169.0	174.8	180.6	186.4	192.3	198.1	203.9
65	108.2	114.2	120.2	126.2	132.2	138.2	144.2	150.2	156.3	162.3	168.3	174.3	180.3	186.3	192.3	198.3	204.3	210.3
66	111.5	117.7	123.9	130.1	136.3	142.5	148.7	154.9	161.1	167.3	173.5	179.7	185.9	192.1	198.3	204.5	210.7	216.9
67	114.9	121.3	127.7	134.1	140.5	146.9	153.3	159.6	166.0	172.4	178.8	185.2	191.6	198.0	204.3	210.7	217.1	223.5
68	118.4	125.0	131.6	138.1	144.7	151.3	157.9	164.4	171.0	177.6	184.2	190.7	197.3	203.9	210.5	217.1	223.6	230.2
69	121.9	128.7	135.4	142.2	149.0	155.8	162.5	169.3	176.1	182.9	189.6	196.4	203.2	209.9	216.7	223.5	230.3	237.0
70	125.5	132.4	139.4	146.4	153.3	160.3	167.3	174.3	181.2	188.2	195.2	202.1	209.1	216.1	223.0	230.0	237.0	244.0
71	129.1	136.2	143.4	150.6	157.8	164.9	172.1	179.3	186.4	193.6	200.8	208.0	215.1	222.3	229.5	236.6	243.8	251.0
72	132.7	140.1	147.5	154.9	162.2	169.6	177.0	184.4	191.7	199.1	206.5	213.8	221.2	228.6	236.0	243.3	250.7	258.1
73	136.4	144.0	151.6	159.2	166.8	174.3	181.9	189.5	197.1	204.7	212.3	219.8	227.4	235.0	242.6	250.2	257.7	265.3
74	140.2	148.0	155.8	163.6	171.4	179.2	186.9	194.7	202.5	210.3	218.1	225.9	233.7	241.5	249.3	257.1	264.8	272.6
75	144.0	152.0	160.0	168.0	176.0	184.0	192.0	200.0	208.0	216.0	224.0	232.0	240.0	248.0	256.0	264.0	272.0	280.0
76	147.9	156.1	164.3	172.5	180.8	189.0	197.2	205.4	213.6	221.8	230.1	238.3	246.5	254.7	262.9	271.1	279.4	287.6

$$BMI = \frac{Weight \times 703}{Height \times Height}$$

Height: Inches
Weight: Pounds

ideal Index: 18.0-24.9
Overweight: 25.0-29.9
Underweight: below 18.0
Obesity: 30.0 and over

Figure 19. Body Mass Index

If one's BMI is below 18, he is considered underweight. If it is between 18 and 24.9, his weight is within the normal range. If it is between 25 and 29.9, he is considered overweight. If it is over 30, he is obese!

Of course, BMI simply provides a rough idea about the status of body weight. There is no such thing that one measurement that truly fits everyone. On the one hand, some people have a heavier skeleton and a higher BMI, even if their bodies appear very thin. On the other hand, some have a thin skeleton and a lower BMI, even if their bodies are actually full and large.

Based on my observation, one is overweight when he is

standing, and finds his abdomen "popping out" or "sticking out" ahead of his chest wall. Or, when he lies down on his back, and finds his abdomen is flat or rises from the plateau of his chest wall.

Chapter 3: Obesity and Health

DIRECT IMPACT OF BECOMING OVERWEIGHT OR OBESE

In January 2005, the US Centers of Diseases Control and Prevention reported that excess weight/obesity had caused about 365,000 deaths during the past year. The report ranked excess weight/obesity as No. 2 among the nation's leading causes of preventable deaths. Three months later, the CDC retracted the January report. It declared that the normal range of BMI (Body Mass Index) should be upward adjusted, because today's American population is "eating better" and "naturally" getting heavier. As the result of the adjustment, the CDC reduced the number of deaths caused by excess weight from 365,000 to 25,814. It even suggested that some people who were slightly overweight were healthier than those who were of a normal weight.

The adjustment quickly offered some relief to those people who were overweight or obese. However, it invited criticism from those people who had always believed that excess weight causes health hazards. Amid the confusion and criticism, the CDC chief made another correction on the earlier CDC report. She acknowledged that its study was flawed. She reiterated that being overweight was not healthy.

I indeed agree that excess weight and obesity may not directly cause so many deaths each year. However, severe medical disorders caused by excess weight, really kill many people every year. Those deaths could have been prevented!

Let us just list a few problems that are directly linked to being overweight. In Chapter 6 of this book and the accompanied reading list, there are studies about the risks involved.

Airway Obstruction

As mentioned, the airway is a very critical concern for anesthesiologists. As soon as an overweight or obese person loses consciousness, while sleeping on his back, the muscles of his mouth and neck relax. This allows the floor of his mouth along with his tongue and epiglottis to fall toward the back of his pharynx.

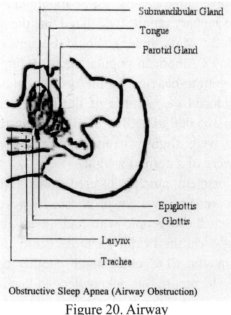

Obstructive Sleep Apnea (Airway Obstruction)

Figure 20. Airway

The *pharynx* is the area of our throat, through which we swallow down the food and fluids into the esophagus. The epiglottis is a lid behind the tongue. Glottis is the entrance of the larynx or airway leading to the trachea. The epiglottis covers the glottis when the tongue moves toward the back of the mouth as a person swallows and sends the food or fluids down into the esophagus. The movement of the epiglottis helps prevent the food or fluids going into the larynx, the trachea, the bronchus, and the alveolus. Alveolus is the terminal chamber of the lungs where the circulating blood takes in oxygen, and then releases carbon dioxide. When we breathe in, a large amount of oxygen, with a smaller amount of other gases, is moved into the alveolus for blood circulation. When we breathe out, carbon dioxide, with a smaller amount of other gases, is moved out of the body.

While we are awake, the epiglottis covers the glottis only when we swallow. Otherwise, it stays away from the glottis, allowing for an open airway. When we are asleep, the tongue and epiglottis usually stay away from the larynx and keep the airway open. As we get older, the muscles of the floor of the mouth become much relaxed. When we are asleep, those muscles and the tongue fall backward toward the pharynx; the epiglottis covers the epiglottis; and the airway is closed. When our airway stays closed, gases are not moved in or out of our lungs, as long as we are sleeping. Then oxygen is in short supply, and carbon dioxide is building up, in both inside the alveoli (the pleural of alveolus) and in blood circulation.

There are two sensors in the respiratory (breathing) center of the brain. One detects the level of carbon dioxide in blood circulation. The other detects the level of oxygen.

When the level of carbon dioxide in the blood circulation is built up, the sensor for carbon dioxide stimulates our respiratory center, which wakes us up to catch breath again. You may have observed that your parents or grandparents gradually breathed more slowly and eventually stopped breathing, after they went to sleep. Suddenly, they would wake up a little bit to catch their breath again. This cycle becomes a pattern for those who suffer sleep apnea or sleep airway obstruction. This happens more often in people who are overweight or obese.

In the case of chronic obstructive pulmonary (lung) disease (COPD) or emphysema, the lungs stay expanded, with little ability to shrink back to their normal or resting volume. A patient with either of these diseases has chronically built up the level of carbon dioxide in his blood circulation. The censor for the level of carbon dioxide in his blood circulation becomes insensitive. When he has airway obstruction, his respiratory center has to depend on the stimulation from the censor for the level of oxygen in the blood circulation. He tends to develop hypoxia (lack of oxygen). However, he is not supposed to receive too much oxygen, or he will stop breathing when his oxygen censor is satisfied, thus ending the stimulation of his respiratory center.

One important reminder is that those who suffer sleep apnea (airway obstruction) should be cautious about taking alcohol and medications, such as sleep pills and tranquilizers. Alcohol and the medications can make the respiratory center less sensitive to the stimulation from higher level of carbon dioxide and/or lower level of oxygen. They may not wake up soon enough to catch their breath and respond to immediate danger. They may have a higher risk of aspirating the content of acid reflux, which would develop inflammation and infection of their respiratory tract, including pneumonia and lung abscess (severely infected with production of pus).

To help cut down the risks of airway obstruction and aspiration, an overweight or obese person would be wise to sleep on his side. Bear it in mind, though, that this would not prevent all risks. Studies show that reducing weight to normal helped reducing the risks of both airway obstruction and a GERD. (To be discussed later.)

Shortness Of Breath

People who are overweight or obese have a higher probability of storing more fat in their abdomen. The fat tends to push against their chest and back. That can cause short of breath, because the lungs have less room to expand, as they take in a breath. The fat may also push out against their abdominal wall to cause umbilical hernia (a hernia as a result of a weakened belly button); and/or ventral hernia (a hernia as a result of separation of a pair of muscles running between the breast bone and the pubic bone).

Just think about this. You are asked to carry a pair of 5-pound dumbbells and walk around for 30 minutes. Don't you keep asking if the time is up, because you are getting a little bit of shortness of breath? How do you feel when you have to carry ten pounds of fat 24 hours a day and seven days a week?

Burden To The Heart

The heart is a pump. It is responsible for driving the blood throughout the body, circulating a supply of oxygen and nutrition to tissues and organs, while removing carbon dioxide and wastes from them. In people who are overweight or obese, tissues require more blood in circulation, thus increasing the burden to their heart. The heart has to beat faster to pump out more blood into the circulation. The increased burden on their heart can contribute to the development of diseases of their heart.

Weakening The Joints

The joints are important components in body movement. They can bear only a certain amount of pressure or weight. The overweight or obese people place more pressure or weight on their joints, especially the hip, knee, ankle, and foot joints. After bearing so much more pressure, their joints may become chronically injured and damaged over time. Joints can move out of alignment and eventually become arthritic and degenerated.

Hurting The Back

The back is a set of bones, ligaments, and muscles alongside the spine. Movement of the back allows us to assume different positions without having to move our entire body around. When we move our back beyond its limits or against certain structures, we may suffer back injuries. People who are overweight or obese build up a mass of fat in their abdomen. They cannot move as freely as they could before the weight gain. This may cause them to move their back beyond its limits, and experience a catch of pain to their back. For example, when they bend over, the fat mass of their abdomen has to push backward against their back that causes pain to their back muscles.

Risk Of Work-Related Injuries

As a result of obesity, those people have a higher risk of injury both inside and outside their workplace. A recent study in the April 23, 2007 issue of Achieves of Internal Medicine, conducted by a group of researchers of Duke University, found that obese workers, as compared to their thin coworkers, filed twice the number of workers' compensation claims, accrued seven times higher medical costs, and lost 13 times more days of work as a result of their work-related injuries. Obesity is a problem that costs both medically and economically. It is not only a personal issue but also an employer's nightmare.

THE INDIRECT IMPACT OF EXCESS WEIGHT AND OBESITY

Other than the immediate impacts that I mentioned in the previous section, many studies have shown a relationship between obesity and other diseases.

Gastroesophageal Reflux Diseases (GERD) & Cancer

During recent decades, we have observed a rise in both obesity and cases of acid reflux. GERD, a result of higher pressure inside the abdomen, pushes gastric juice (acid) up into esophagus. Has the gastric acid become more acidic because of the high-carbohydrate, low-fat diet, which has caused weight gain? Although we still do not know why overweight and obese people have a higher risk for GERD, the severity of the disease deserves our serious attention.

A study published in the August 2, 2005 issue of Annals

of Internal Medicine, conducted by Dr. Howard Hampel and his associate, reviewed all related, published studies, between 1966 and October 2004. Six out of nine published studies found a significant relationship between body mass index and the symptoms of gastroesophageal reflux disease. Six out of seven studies linked body mass index to erosive esophagitis (inflammation with wearing down of the esophagus). Six out of seven studies found a strong relationship between the body mass index and the esophageal adenocarcinoma (a type of cancer of the esophagus). Four out of six studies found a significant relationship between body mass index and the gastric cardia adenocarcinoma (a type of cancer of the upper part of the stomach).

People with a BMI between 25 and 29.9 (overweight) had a 1.43 greater risk of having symptoms of GERD than those with a BMI below 25. For those whose BMI were 30 and over (obese), the risk was 1.94 times higher. In the same comparison, the risk of esophageal adenocarcinoma (cancer) was 1.52 times for those with a BMI between 25 and 30; 2.78 times for those with a BMI over 30.

Diabetes Mellitus (DM) And Insulin Resistance

Studies have repeatedly named excess weight and obesity as a cause of diabetes mellitus (DM) and its complications including insulin resistance. However, I personally believe that both diabetes mellitus and obesity are just part of the growing diseases resulting from hyperglycemia after eating. Weight gain and obesity are a critical warning that those people should check the pattern of their blood glucose levels. Their blood glucose levels are likely higher than a healthy limit, and will cause diabetes mellitus sooner or later.

Based on the current "flawed" criteria for diabetes mellitus; not all the people who are overweight or obese are diagnosed with diabetes mellitus (DM). However, the people who are of a normal weight or underweight are not necessarily healthy, nor are they exempt from developing diabetes mellitus (DM). Those with normal weight or underweight may be taking a daily amount of calories that are just enough for their daily activities, and not enough for gaining weight. However, if they take in most of the calories from carbohydrates, their blood glucose level may be higher than normal, sufficiently beyond healthy limits to cause diabetes mellitus.

Cardiovascular And Other Diseases

In addition to its connection to diabetes mellitus, many studies have shown a link between excess weight and hypertension, coronary heart disease, stroke, dementia, inflammatory diseases, cancers, as well as other serious ailments. When we carefully explore the common cause(s) shared by excess weight/obesity and these diseases, we can trace all of them to hyperglycemia. A case of hyperglycemia should not be limited to the higher-than-normal blood glucose levels of fasting or the two-hour period after a glucose tolerance challenge. Rather, hyperglycemia should be identified as such any time the blood glucose level is higher than a "healthy" level, say 150 mg%.

By the way, "fat but fit" has been an issue in debate for recent decades. Some scientists, especially those who are overweight or obese themselves, would strongly assert that as long as a person stays active physically, excess weight should not increase his health risks. After reading both the direct and indirect impacts on our health by obesity, few would agree that it is safe to believe "fat but fit."

A recent research report by Amy R. Weinstein, M.D. and her research team, was published in the April 28, 2008 issue of Archives of Internal Medicine. Despite the fact that physical activity helped reduce the risk of coronary heart disease for women, those who were obese or overweight had the risk of coronary heart disease at 1.87 or 1.54 respectively, as compared to those whose weight was normal (with the risk of coronary heart disease at 1.00). Among those who were inactive physically, as compared to those whose weight were normal and active physically; the obese-inactive women had the risk of coronary heart disease at 2.53; the overweight-inactive at 1.88; and the normal weight-inactive at 1.08.

Chapter 4: Weight Gain And Diet

WE ARE WHAT WE EAT!

We need the calories in foods to keep our body running constantly. Depending on the type and amount of foods we eat and the level of our activities, we gain or lose weight. On the one hand, we gain weight because the amount of calories we take in is more than we need to support our activities. We save the excessive calories in our body fat. On the other hand, we lose weight because the amount of calories we take in is less than we need to support our activities. To make up for the shortage of the required calories, we convert our body fat into calories for our activities. This is a simple principle in bookkeeping.

We can measure the output of power an engine produces when it burns fuel into carbon dioxide and water. But we cannot easily measure the amount of energy that our body machine generates by burning the foods into carbon dioxide and water. Instead, by measuring the amount of heat that our body generates while converting the foods into energy, carbon dioxide, and water, we can determine the amount of energy for each type of food.

In the laboratory, we use a bomb calorimeter to measure the amount of heat each type of food generates in the process of converting the food into water and carbon dioxide. Each gram of carbohydrate foods generates about 4 kilocalories of heat when it is converted into water and carbon dioxide. About 9 kilocalories come from each gram of fat; about 4 kilocalories are from each gram of protein; and about 7 kilocalories are from each gram of alcohol. (Source: Textbook of Biochemistry with Clinical Correlation, by Thomas M. Devlin, 5th Edition, 2002.)

A biochemical process of the body, Tricarboxylic Acid Cycle or TCA Cycle, uses acetyl coenzyme A (acetyl CoA), for producing energy. (Please see Figure 5 Metabolism of the Nutrients, Chapter 1.) There are different amounts of energy loss in processing carbohydrate, protein, and fat into acetyl CoA for TCA Cycle. Typically, using glucose from carbohydrates is very efficient in producing energy. However, there is a loss of one calorie for every gram of protein, when it is converted for producing energy; one gram of protein actually produces only three calories of energy. This is the reason why, in trading carbohydrates for proteins in diet, gram for gram, the protein-rich diet can help lose weight.

WHAT IS GOING ON WITH OUR DIETARY RECOMMENDATIONS?

Over the past decades, the United States government has issued food pyramids and "Percent Daily Values" for us. It meant to help us eat "healthy foods" and keep us from being ill with serious diseases such as coronary heart disease, stroke, and et cetera.

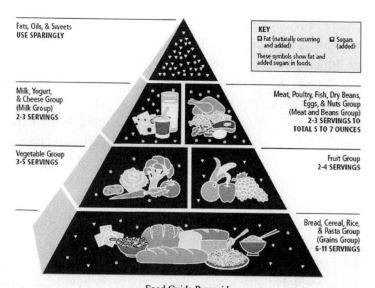

Food Guide Pyramid

Source: Dietary Guidelines for Americans. March 2003. Center for Nutrition Policy and Promotion. US Department of Agriculture. Home and Garden Bulletin No. 267-3

Figure 21. Food Guide Pyramid

"Percent Daily Values" are based on a 2,000-caloire diet. Your daily values may vary higher or lower depending on your calories needs:

	Calories	2,000	2,500
Total Fat	Less than	65 g	80 g
Saturated Fat	Less than	20 g	20 g
Cholesterol	Less than	200 mg	300 mg
Sodium	Less than	2,400 mg	2,400 mg
Total Carbohydrates		300 g	375 g
Dietary Fiber		25 g	30 g

Figure 22. Percent Daily Values

If these dietary recommendations work, why does our population keep growing heavier? Excess weight and obesity are found not only in the adult population, but also among the young. More typically adult diseases are appearing in our children, at younger ages each year. Shouldn't the government, and both the medical and nutritional experts stop and recheck their concepts about the relationship between nutrition and health, then overhaul their dietary recommendations?

Lately, more authorities are calling for a decrease in sugar consumption, particularly for children. But most people do not understand the link between overeating and satiety. They do not know the importance of glycemic index and glycemic load. They falsely believe that sugar is the only cause for weight gain and obesity. They should know that foods with a high amount of polysaccharide, disaccharides, and monosaccharide are an excellent source of blood glucose, too. Eating too many carbohydrates raises one's risk of serious weight gain and illness.

As of this writing, the dietary recommendations from our government as well as the medical and nutritional experts still strongly recommend eating carbohydrate-rich foods. By looking at the recent food pyramid, they are still telling us we should eat more grains, vegetables, and fruits. At the same time, we should eat less cholesterol and fewer fats, especially the saturated fats.

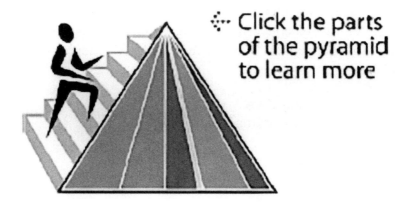

❖ Click the parts of the pyramid to learn more

Color Representation
Chocolate: Grains. **Green**: Vegetables, **Dark Red**: Fruits,
Corn Flower Blue: Milk Group, **Dark Blue**: Meat and Beans.
Source: United States Department of Agriculture
http://www.mypyramid.gov/pyramid/index.html

Figure 23. Inside the Pyramid

These recommendations continue to promote the wrong concepts: *Carbohydrates carry less than a half of the calories of fats, gram for gram. Carbohydrates are good for us, and allows us to eat more in volume "without' harmful effects. Fats are dangerous, because they cause arteriosclerosis, heart diseases, and et cetera. Weight gain and obesity are results of overeating fats and exercising too little.*

The flawed concepts totally ignore the need to recognize satiety, to control our appetite and the volume of our meal. The experts also neglect the relationship between the quality and amount of carbohydrate and its level of blood glucose (sugar). The worst of all, they fail to recognize a vicious cycle that eating more carbohydrates causes a "Sweet Roller Coaster" of hyperglycemia, hypoglycemia, and hunger for more carbohydrates. That "Sweet Roller Coaster" is pre-diabetic.

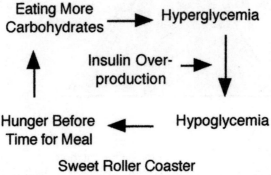

Sweet Roller Coaster

Figure 24. Sweet Roller Coaster

After reviewing the articles listed in Chapter 6 "Carbohydrates Can Kill: A Mountain of Evidence," and the online reading list, we cannot help but ask, "What is going on with dietary recommendations?"

Chapter 5: A Physician's Personal Experience

I TRIED STAYING HEALTHY

According to my late mother, I had been sick very often in the years before I began school. I had suffered from small illnesses like the common cold to big ones such as dysentery. During the 1940s, my birthplace, Taiwan, had experienced a serious shortage of food and health care.

My mother was a very smart lady and an excellent observer. She taught me many things from the time I was a very young child. When I caught a cold, she told me to avoid eating sweets and banana. She said that these foods would make me cough harder with more phlegm. She promised me that she would give me more of my favorite sweets after I recovered from the cold.

For most of my life my skin has been very sensitive and frequently itchy. Since I was a small child, I had scratched my limbs and body. This was especially true when insects touched or bit my skin. Often I scratched so hard that I broke my skin to bleeding. When my mother saw my infected wounds, she warned that eating too many sweets made wounds much harder to heal. My mother said that personal preference for sweets gave me a weaker "body constitution." I did not know what she meant at that time.

However, I started to stay away from sweets like candies and cookies. As a child, I did not know all carbohydrates should be considered "sweets" too. Even my mother did not realize that either.

After entering the grade school, I became healthier. I started eating better; I ate everything on the table. When I attended middle school, I was fairly healthy, except for an occasional cold, headache or neck ache. During my high school years and until I began pharmacy school, I got headaches more often, about three times a week. I was badly sick once a week from the severe ones. I had to take aspirin and sleep in a dark room to get over them.

In the first lecture of my pharmacy school, the dean told us that most medications were poisonous. He explained that there is a therapeutic dosage, a toxic dosage, and a lethal dosage for most medications. In addition, most medications produce side effects or complications. I said to myself, "I will make the best efforts in keeping up my health and staying away from medications."

In 1965, I got my pharmacy degree and started military service. I felt great and I had become much healthier. I went to Japan in 1967 and began my medical education at Nagasaki University School of Medicine. I studied very hard, because I found that I was very interested in medicine. During the next four years in the medical school, my wife and I welcomed two more babies into our family. In addition to studying medicine, I had to help my wife take care of our three children. I also made good grades at school. At all time, I had maintained my weight at around 145 pounds. Although I was so busy, I felt great and very energetic. I needed only 5-6 hours of sleep a night. In the spring of 1971, I graduated from the medical school as one of the top five of my class. Until the end of June that year, I interned in surgery at the Nagasaki City Hospital.

I came to the United States in the summer of 1971. For the next three years, in my internship and residency, I put in about 80 hours of work a week. However, my weight stayed at about 145 pounds during that time. I was in very good health but I did suffer headaches regularly, especially after having worked a long shift. Headaches were a result of breathing anesthetic gases so often in the operating room. I have never wanted or intended to be an anesthesiologist. I was looking forward to the start of an obstetrics/gynecology residency.

In the spring of 1974, as I was finishing my anesthesiology residency, the chairman of the department offered me a fellowship in the department. He wanted me to help set up a pain clinic with acupuncture practice. I declined. When the chairman learned of my applications for the obstetrics/gynecology residency programs, he refused to write the required recommendation letters. Instead, he asked me to enter private practice, in lieu of accepting the fellowship with the department.

Shortly thereafter I started private practice as an anesthesiologist. I continued working hard. The stress from taking care of sick patients and the ongoing exposure to anesthetic gases still gave me regular headaches. Thanks to my self-administered acupuncture, I suffered fewer headaches years later. In the meantime, I found my weight had slowly increased. At the turn of 1980, I was 155 pounds. However, I was still very healthy and hardly ever got sick.

The biggest advantage in being a physician is the

capacity to observe myself carefully at all times. I have never let my emotions influence my medical judgment on any patient's conditions, including my own. During that time, a nurse sent by the life insurance company, conducted a series of laboratory tests and checked my blood pressure once a year. The results assured me that I was in good health. Until 1990, my lab results had remained within the normal limits. My blood pressure was always around 120/60 mmHg. However, I noticed that I have been slowly gaining weight since the early 1980s.

By the end of August 1997, I had grown very tired of my practice in anesthesiology. Due to hospital ownership changes, I was forced to abandon my solo practice to participate in the group practice contracted by the only local hospital. I was very unhappy about the changes. I retired from practicing in anesthesiology at the hospital. It was an excellent decision that helped me ease the pressure of 23 years. I had never enjoyed practicing in anesthesiology.

Of course, I did not want to just idle myself and settle into my early retirement. I needed to work for keeping me busy and paying my bills. In the meantime, I had a big adjustment in my daily routines too. I no longer had to rush to the operating room early in the morning. Neither did I have to walk from floor to floor to check patients, before and after surgeries. Nor did I have to help nurses transport patients between the recovery room and the operating room. Instead, I increased my office hours for a full time practice in pain management. I just went to my office in the morning based on the hours of my patients' appointments. My patients heard my fast footsteps between the examining rooms in the 1,350 square-foot office. They knew that I was so busy. They thought that all of the walking I did in the office must have equaled a good deal of daily exercise, and I thought so, too. Before retiring from practicing in anesthesiology at hospital, I used to spend time working many afternoons around my house, when I had finished my daily schedule. I mowed grass, spread fertilizers, cut trees, and fixed things around the house. After retirement, while practicing exclusively at my office, I not only had to spend more time seeing my patients, but I also had to bring home office paperwork for dictation of patient records. More often than not I had to work several hours in the evening at home. I did not have much free time for working around my house as I had before.

SUMMER 2002

I Had Problems!

In the early afternoon of June 9, 2002, I came home from a regular weekend shopping trip. I had bought a new electronic blood pressure machine for replacing the old one in my office. It was a nice machine with automatic pump-up of the cuff. It had a LCD window, which displayed digital readings of the blood pressure and the pulse rate. The cuff was a regular adult size. The one end of the cuff was threaded through a metal buckle, which was fixed to the other end of the cuff.

To satisfy my curiosity about the quality of the new machine, I quickly opened the package and set it up. I sat down and slipped my right arm into the blood pressure cuff, then properly wrapped the cuff around my right upper arm, just about one inch above my right elbow. I made sure that the cuff was not too tight or loose, then pushed the button of the machine to start.

After the cuff went up tightly, and then slowly loosened its grip over my right upper arm, I could hardly wait to take a glance at the numbers, displayed inside the window. To my disbelief, my blood pressure was 205/63 mmHg, with the pulse rate at 68 per minute. I shouted to myself, "I cannot believe this machine! There must be something wrong with this machine. I just cannot believe this machine!"

But, how could I dismiss the machine just because I did not like the reading? For the sake of my own health, and having a good blood pressure machine for my office, I knew I had to find out what was going on.

As if I were examining a patient, I began to do some checking for the accuracy of the blood pressure machine. I switched the cuff onto my left arm; and found the reading was 209/62 with the pulse rate at 66 per minute. It was awful and scary. Again, I switched the cuff back to my right arm, repeated the blood pressure reading about once every minute. About ten minutes after I had first begun to play with the machine, the reading was 194/58, with the pulse rate at 72 per minute. My blood pressure was still too high! After repeatedly checking my blood pressure, for a total of twenty minutes, the last reading of my blood pressure from my right arm, was 195/60 with 60 per minute for the pulse. There was no doubt that I had problem with my blood pressure.

I began to recall what I had noticed about my physical changes in recent time. I indeed had had several symptoms that justified my concerns over the last few years. I had been experiencing a crushing ache in the left side of my upper back every time I got excited or when I rushed myself a little bit or climbed the staircase one time after another. It had happened when I started to work in my yard on a mild, sunny day or when I took a slightly fast walk with my dogs in the city park after a regular visit to my parents' grave on a weekend or when I was occasionally waiting for the end of bidding on a much-wanted eBay item. However, the crushing pain and ache always went away in less than a minute, after I stopped rushing myself or got over my anxiety. I even had increasingly felt the same discomfort or ache in my left upper back in the middle of intimacy during the past year or so. It alarmed me very much, but I had never told my wife.

More often during the last couple of years, I had often heard a rushing noise in my ears, especially on my left one, that matched my heartbeat. The noise bothered me the most when I lay down in bed at night.

I knew that many men my age might consider it a minor problem, but I was uneasy and aware of a noticeable weakening of my manhood. I believed that I did not have any symptoms to suggest that I was diabetic. I always loved to drink a lot of fluids, purposely for body fluid replacement, and sometimes because I was somewhat thirty. Thirst can be a symptom of diabetes mellitus (DM). However, the worsening impotence was certainly one more reason for concern. This had been one more "alarm" that I must give this my immediate attention before my time ran out. I kept reminding myself of more and different occasions that brought on that suspicious, serious ache in my left upper back.

The big gap between my systolic blood pressure (the upper number of the reading) and diastolic blood pressure (the lower number of the reading) suggested that I might be ill with either aortic insufficiency, arteriosclerosis, or both. The aorta is the largest artery of our body. The ascending aorta is the first part of the aorta next to the left side of the heart. Between the left lower chamber of the heart (the left ventricle) and the ascending aorta, there is a pair of aortic valves. The aortic valves are open, when the heart pushes out the blood into the ascending aorta, and then to the periphery of our body. The blood pressure at this point is the systolic one. To make a perfect delivery of the blood to the periphery, the aortic valves must be closed, after the heart empties its load and before it relaxes itself. At this point, the blood pressure is the diastolic one. The aortic valves prevent the blood from flowing backward into the left ventricle. It is most important that when the aortic valves close, the blood flows into the coronary arteries and supplies oxygen and nutrients to the heart.

Aortic insufficiency happens when the aortic valves are defective. In such a case, the aortic valves fail to completely close the opening between the left ventricle and the ascending aorta; then the blood flows backward into the left ventricle. As a result, the diastolic blood pressure falls. The blood flow to the coronary arteries is reduced, and not enough circulates through the heart. In the meantime, the backward flow of the blood into the heart makes the heart work harder the next time it pumps out the blood. The heart has more blood in its chamber than it did before. It cannot rest properly or re-supply itself with enough oxygen and nutrients. The heart is in pain. This is one kind of heart attack.

So, I took out my stethoscope and listened to my heart sounds. There was no murmur from any of the valves of my heart. Good, I didn't seem to have a problem with my heart valves, but I could not guarantee that I would not have one. If I let the situation go unattended it could get worse. Next, I placed the stethoscope on each side of my neck, and listened to the carotid arteries. Well, I did not hear any noise in either one that would suggest that I might have carotid stenosis (narrowing of a carotid artery.) Carotid stenosis is a result of arteriosclerosis. However, I heard humming off and on in my right carotid artery. By the time I had relaxed, the humming went away completely.

Arteriosclerosis is the hardening of the arteries by build-up of cholesterol plaques onto the inner layer of the arteries. It makes the wall of the arteries unable to stretch. That means the arteries cannot extend to make room for sufficient blood flow, causing a high systolic blood pressure. I thought the rushing noises matched my heartbeat and the occasional humming that I heard in my ears indicated a dangerously high systolic blood pressure. I believed that I had arteriosclerosis, hypertension, and likely angina. Worst of all, with the blood pressure readings at that time, I feared I could suffer a deadly heart attack, stroke, or both at any moment.

I remembered reading an article in a magazine. It was an interview that a well-known chest surgeon had agreed to give, a week or so after he turned 50 years of age and shortly before his death. He was chief of the chest surgery department of a top university hospital. He started smoking when he was 5 years old. He became a heavy smoker as he grew up. He continued to smoke until his death from an inoperable lung cancer. His secretary was often embarrassed when his patients asked her about the ashtray, full of cigarette butts, on his desk. What did his patients think about this doctor who urged them to stop smoking because of their lung cancers, while he smoked heavily himself?

I also remembered a story from another magazine. A patient wrote the editor of her local newspaper, expressing her surprise at having learned her doctor was ill when she called his office for an appointment. We all should know that doctors are human beings, just like anyone else. Still, many patients expect their doctors should stay healthy all the time.

Now I was very worried. "What if I have a crippling heart attack or stroke?" I asked myself. I would have to sit in a wheelchair for the rest of my life. My wife and I had worked hard for all our life. We had not yet had a chance to see the world outside our area. I had just entered the 61st year of my life, and I was looking forward to spending time with my wife, traveling around the country and the world. Now I saw myself sitting in a wheelchair, with my wife needing to push me everywhere we went. I said to myself, " Oh, no! I cannot let it happen! Never! I got to do something now!"

I had seldom weighed myself over the past decades. I was fairly thin in my young adulthood. I had never had to watch what I was eating and I had not gained much weight either. I thought that I had no reason to worry about my weight. I got on the scale a few times during the last few years, before retiring from my practice at the hospital in August 1997. I would stop by the surgical recovery room to check my patients, before leaving the hospital. I would step on a scale on my way out. I saw the scale reading go from 170 pounds to 175, 180, then 185. It was finally 190 pounds when I was on that scale last time. I had never taken the reading seriously, because I was fully dressed. I had a few medical tools, such as a stethoscope and a beeper, in the pocket of my dress jack. Besides, I was on the scale with my shoes on.

As soon as I found out that I had such a high blood pressure on that June afternoon in 2002, I took off all my clothing but underpants. I went to the scale in my bathroom. My goodness! I was 186 pounds! Based on the BMI, I was now at least 26 pounds overweight. It was unacceptable! In addition to my weight gain, which might likely have caused the high blood pressure, I had also struggled with obstructive sleep apnea for at least a couple of decades. I knew that I had an increasing number of nightmares as I gained more weight. Please see AIRWAY OBSTRUCTION in Chapter 3.

A person who suffers obstructive sleep apnea likely suffers snoring too. Mouth breathing dries up his mouth; makes his epiglottis (the cover for the glottis) stick to his glottis (the entrance to the larynx), and hard to open his airway at times. During a period of a few seconds, when his breathing center wakes him up, he would experience a life-threatening struggle. He could dream of being chased and unable to run away. Sometimes, he could be waking up but unable to move his body or make a scream. He could feel that he would be all right if he could just make a loud shout. What a nightmare!

My wife had to shake me and wake me up. She was worried about me. I was worried too! I wondered whether I could wake up to catch my breath in time or if I would suffer a heart attack, or a sudden death!

For more than a couple of decades, I have adapted to sleeping on my sides. When I sleep like this, my tongue, the bulk of relaxed muscles in my neck and the floor of my mouth do not fall back toward the back of my mouth. The epiglottis does not close up my glottis and airway. During the sleep, it is difficult to stop me from rolling myself from one side to the other, or sleeping on my back. So, when I ended up sleeping on my back, I would have a risk of obstructing my airway. The stress from the repeated buildup of carbon dioxide and the decrease of oxygen in my blood may well have raised the blood pressure. That could also have caused irregular heart rhythms, such as atrial fibrillation (AF) and premature ventricular contraction (PVC)!

For years, cardiologists used the crease(s) in a person's earlobe(s) as an indication of coronary heart disease. During the recent years, we have questioned the credibility of such observation. But, I thought that creases in a person's earlobes are a result of his habit of sleeping on his side, probably because he could breathe better in this position, especially if he is overweight or obese.

Improving My Cardiovascular System Without Medications?

After finding out about my problems with high blood pressure, I was really worried. My mind was running fast in every direction. I had worked for most of my life to take care of my patients both in the operating room and in my office. I had saved many patients on the verge of death. Now it would be a joke of my life and tarnish my medical career, if I suffered a sudden death. I said to myself, "No, I cannot let it happen. Yes, I must do something now!"

Then, what should I do now? I thought about seeing one of my colleagues, either an internist or family physician. But, what was he or she going to do for me? Without a doubt, I knew that I would be prescribed diuretics to ease the fluids in my body, especially inside my blood vessels. Diuretics would shrink my blood volume, bring down my blood pressure, and reduce the workload of my heart. Bringing down my blood pressure was important to preventing a heart attack or stroke. I began to worry that concentrating the blood inside my vessels would increase the thickness of the blood flow. That would increase the chances of blood clotting. That could increase the chances of heart attack and stroke.

Besides, I would be given other medications such as a calcium-channel blocker or a beta-blocker. These medications would help reduce the work of my heart. But these medications would reduce my stamina even further. I would have to carefully watch my physical activities to avoid over-burdening my heart. I would likely have to take a baby aspirin or a kind of blood-thinner every day. Aspirin and blood thinners would prevent blood clotting from causing heart attack, stroke, or both.

As I mentioned before, I remembered that introductory lecture from the dean of my pharmacy school. He said, "Most, if not all, medications are poisonous. Other than their side effects and complications, they can be therapeutic, toxic, and deadly, depending on their dosage."

Pledge The Rest Of My Life For Experiment

I kept thinking about the advantages and disadvantages of the options for treating my cardiovascular problems. As a physician who had practiced nearly thirty years, I completely trusted my own knowledge and clinical experience. They were good enough to take care of my patients so they should be good enough to take care of me. But I could be powerless in handling my own care if I were in an emergency situation.

Of course, this kind of consideration is not for someone who is not a medical professional. And, please do not try my way at your home! You must always consult a physician for proper care.

So, I decided to be my own physician, as I had always been during the last three decades. I have never doubted that, other than the air we breathe, the radiation we receive, and the magnetic field we live in 24/7, we are what we eat. I was going to do some experiments on myself to see what would happen. If my experiments succeeded, I would have made a major accomplishment to further validate my trip (or life) to this world.

I understood that lowering my systolic blood pressure must be the top priority. In the past, I had purposely increased my daily fluid consumption in order to support optimal physical functions. I usually drank 12 to 14 of 8-ounce glasses of fluids on a typical day. That probably overloaded my blood vessels and caused high blood pressure. Now I was going to carefully reduce my fluid consumption and salt intake. I had thought that salt prevented water inside the body from going out in the urine. I watched the color of my urine for its concentration to avoid dehydration and kept my daily fluid consumption to eight of 8-ounce glasses a day. I also asked my wife to add no salt to my food. I found that reducing the intake of fluids and salt was not difficult. My trips to the restroom were noticeably less frequent. The color of my urine was much concentrated.

Although I could not exercise or work very much in my yard, I started to take a slow walk for a block or so. I loved to be outside walking in the neighborhood. But I did not like the disadvantages, such as walking in the rain or in the dark. I disliked having to get dressed properly for those walks, too. Not long after I began to walk for exercise I decided to walk indoors instead. At the beginning, I walked back and forth between rooms of my house every evening. I tried to walk slowly to avoid triggering pain in my left upper back. I thought the pain was a result of high pressure of the blood flow, which was pumped out of my heart against the wall of my aorta. I thought it could be from a case of exertional angina (angina in hard work.)

In the meantime, I checked my blood pressure several times a day to see if it had started to come down. I also got on the scale a few times a day, hoping to see that needle had started moving to the left. Eventually, I walked inside the house once every day for the rest of June 2002. I felt that I needed to increase the physical challenge. I began to work outside in my yard. I mowed the almost-an-acre yard with a push-mower, some with a riding mower. Yes, I should have used only the push-mower for the entire yard work. I felt I must be careful with my heart and blood pressure. I wanted to avoid overtaxing my heart with too much work all at once. For instance, I would experience the crushing pain in my left upper back and shortness of breath, about ten seconds after I began to push the mower. Until this summer, I had thought that the bad experience in mowing grass was brought on by gasoline exhaust irritating my lungs. But, I now knew otherwise.

No-Salt Diet Did Not Help Me At All

By early July 2002, even with the restriction of fluids

73

and a no-salt diet, my systolic blood pressure had not come down. It was still between 180 and 190 mmHg, when I was relaxed. It also was between 200 and 220 mmHg when I was excited or physically active. When the machine registered 220 mmHg, for my systolic blood pressure, I was shocked and very worried.

I knew that I had to be patient with my efforts. That was the only way to bring my health back, if I still did not want to use medications or "toxic" chemicals. So, I thought that I must do more physically.

There are three sets of stairs in my house. One is about 10 feet high between the first and second floor of the main house. It has 16 steps. Another is about 7.5 feet high with 11 steps between the kitchen and a room over the garage. There is an opening between a room over the garage and the second floor of the main house. The third staircase is about 2.5 feet, with five steps next to the opening.

I decided to design another trail with climbing the stairs. I climbed the intermediate flight of stairs from the kitchen to the room over the garage. Then, I walked about thirty feet to the shortest set of the stairs between the room over the garage and the main house. After that, I walked about 15 feet across a room to the longest flight of stairs and then down the stairs. Lastly, I walked across the kitchen to the bottom of the first flight of the stairs to begin the trail again.

We All Must Learn To Listen To Our Body!

The trip upstairs was divided with a walk of 30 feet between the intermediate set and shortest sets of stairs. After climbing five steps and walking another 15 feet, I walked down the longest flight of stairs and across the kitchen.

Walking on the floors between stairs added the time for my heart to recover from the stress by climbing the stairs. Then, I was ready for another climb. This made good sense in terms of cardiovascular rehabilitation.

At the beginning, I took my time on the trail as I went up and down the stairs. I was able to walk two or three laps before having to take a break for one or two minutes. I could feel the ache in my left upper back after climbing the stairs for the third time before taking a break. However, as I later walked a few more laps it felt easier. I initially set a goal of ten exercise laps for each day. I was able to finish ten laps in 25-30 minutes.

By the middle of August 2002, I had increased the number of laps of walking in the trail from 10 to 20. I spent 40 minutes every day doing this exercise. To avoid getting bored while walking, I carried a CD player with a pair of earphones. Listening to classical music on the trail helped me to relax.

By the end of the summer, I had managed to finish 20 laps of walking on the trail. I did not have to take a break on a regular basis in the middle of the exercise, only occasionally as needed. However, I still felt the heavy pressure or ache in the left upper back each time I finished the third or fourth lap. I usually stopped for a minute or so before continuing to my goal. My legs and hips became tight and sore after finishing the first two or three laps. I must admit that I got tired of walking sometimes. I even wondered if I should continue to "punish" myself by doing this hard and boring workout. For a split second, from time to time, I saw the image of myself in a wheelchair, on which my wife was pushing. Oh, no! I told myself that I must continue the project to erase that image from my life!

In the meantime, I weighed myself at least twice a day. Most days, I did so more than three times. To make sure that the reading of the scale was as accurate as possible, I always stripped myself to the underpants. On an ordinary day, I got on the scale before the bedtime and immediately after getting out of the bed in the morning after finishing the restroom business. I called the first reading of my weight in the morning the "empty weight" or, in the medically appropriate term, the "fasting weight." By the end of August 2002, I had been on water restriction, no-salt diet, physical exercise, and yard work for almost three months. But I had not lost weight at all. I was very disappointed and worried, to say the least.

A few times, I thought about giving myself up to the "traditional medicine", which would use chemicals to manipulate my body and change its physiological functions. "No!" I quickly said to myself, "I do not want to use chemicals at the expense of my vital organs, just for the sake of correcting my symptoms. It is very un-physiological!" I must be more patient and must give myself more time, before I make any changes in my experiments.

FALL 2002

Exercise Alone Makes No Improvement

After entering September 2002, I continued trying to lower my blood pressure, and ease the left upper back pain during exertion by carefully limiting fluids I consumed and adding no salt to my food. I still tried to walk 20 laps on the stair-trail each day, at least three out of every four days. I still noticed heavy pressure and occasional aching in my left upper back, after having walked the third or fourth lap. I felt that the condition of my cardiovascular system might be slightly improved, but not very much.

On September 28, 2002, my wife and I flew to Maui, Hawaii. We joined our son, daughter-in-law, and lovely two-year-old grandson there for a weeklong vacation. My wife and I stopped in Hawaii on our way to the United States from Taiwan 31 years ago. We had never been to Hawaii for a pleasure trip.

It was a wonderful vacation that my wife and I really needed. As a solo practicing anesthesiologist, I had never been able to find much time for vacations. After retiring from practicing anesthesiology, but continuing office practice in pain medicine, I was still unable to block a week of my schedule for a vacation.

During the stay in Maui, we went to the beach many times. The beach was a half-mile from the condominium where we stayed. I did not have the stair-trail for exercise there. However, I took the walk to and from the beach for exercise. Sometimes, I helped push my grandson's stroller. I still noticed pressure and aching in my left upper back, shortly after I started to walk. About five minutes later the ache had eased up. As a precaution, I did not go into the water. I stayed on the beach and played in the sand with my grandson.

We also visited shopping centers and a volcano, as well as several gardens and nurseries. My son drove us around more than three-quarters of the coastline. However, we did not walk very much, other than to visit the gardens on the mountain. I did not feel that I got enough of a chance to appreciate any improvement in my physical condition.

We returned home on October 5th, by a red-eye flight. Many passengers were coughing in the cabin. In the evening we arrived home, I became sick. I had developed a bout of upper respiratory infection caused by the contaminated air inside the plane. It began with sneezing and a running nose with clear discharge, for the first three days. On the fourth day, I had a sore throat from the nasal discharge that drained into the back of my nose and mouth (pharynx). Soon after that, I began to cough and produce phlegm, which became yellowish. It took me 10 days to recover from the whole ordeal. I had to prescribe myself amoxicillin for the upper respiratory infection. Even during my sickness, I resumed walking 20 laps of the stair-trail for exercise, at least three of every four days.

By the middle of October, I had found no improvement in my blood pressure. I continued to notice the left upper back pain after having walked the third or fourth lap of the trail. My systolic blood pressures even went up to 220 mmHg, when I was excited or after I had walked three or four laps.

In the meantime, I only saw a very little change in my weight. I was very discouraged! I wanted to be patient and allow myself more time for turning things around. But, what else could I do to lose weight?

Even with my best efforts in controlling the intake of fluids and salt, I really could not see that I was any closer to a solution to my problems. I believed that I needed to look into other options for improving my physical conditions. Dropping my weight would make all of my movements easier. I needed to set a higher level of exercise. I decided that I should probably try to walk 40 laps of the stair-trail in every session.

Just think about this. It was going to be a big difference in exercise for me to carry a pair of five-pound dumbbells during a 30-minute session. I would have to keep the muscles of my body, especially those of my legs, my abdomen, and my back, working harder in carrying the extra weight. To make those muscles work harder, my heart would have to beat faster and pump out more blood in each beat for carrying more oxygen and nutrients to the muscles. The heart would work harder and need more oxygen and nutrients to its own muscles, too.

When my heart needs to work harder and beat faster, it has shorter time between beats. The higher the number of the heart beats, the shorter the time for relaxing the heart. That means it has less time for its coronary arteries to receive blood for its own muscles. If the coronary arteries in my heart have a narrowing or blockage in their passages, my heart muscles will have spasms, cramps, and will even be damaged. For that, I was concerned about the heavy pressure and ache in my left upper back, when I was walking, working in my yard, or excited.

"Now, I weigh at least 26 pounds more than I should," I said to myself. "That extra 26-pound is with me every second, every minute, 24 hours a day, and seven days a week. My heart has to work much harder for the sake of the 26 pounds of extra weight all the time. No wonder I cannot walk or do anything, even just for a while."

I had not been able to reduce my weight, bring down my blood pressure, or ease the pain in my left upper back by drinking less water and having no salt in my foods. I needed to forget about the belief that the salt and water were the main factors for my high blood pressure. Also, I had always doubted the efficiency of exercise for weight loss. I had seen people running around in my neighborhood every day. They were running hard and sweating a lot. Many still carried a big belly. I have also seen people working hard at various construction sites. They appear to work very hard and burn lots of calories every day. However, many of them have big bellies that seem to be getting bigger all the time. Why? Aren't they burning lots of calories during their hard physical work?

Change My Diet

I recalled what I learned from the courses in physiological and nutritional chemistry at pharmacy school. It is not fats or proteins, but the excessive carbohydrates, with the help of insulin that become fat and glycogen stored in the body. It is not efficient to use protein as the source of energy, because of the energy loss.

We are often given the number of calories for each gram of fat, protein, and carbohydrate in foods. That number is measured *in-vitro* or "outside the living body." That means the number of usable calories, for each gram of fat, protein, and carbohydrate, should be different inside our body or *in-vivo* as compared with the measurement in-vitro. A person who consumes a 2000- calorie diet does not always produce 2000 calories inside his body. The state of each individual's digestive system is different from that of other people, to say the least. I have always believed that using the number of calories alone in managing our nutrition is impractical, and does not make sense!

So, I go back to my own history. I knew that if I really

wanted to lose weight in hope of easing the burden on my heart and dropping my blood pressure, I would need to restrict my consumption of carbohydrates, especially those with high glycemic indices. I decided to change my approach in my continuing life experiments. It was on October 12, 2002 that I told my wife of my plan to stop eating breads, cakes, ice cream, noodles, potatoes, and any foods with flours, starch, and sugar. I had not been in love with candies or cookies since I was a youngster, because of my mother's warning. I would have no problem at all to stay away from them. I also stopped drinking juices because of the sugar, which is added by the manufacturers. I had not drunk Coke and other soft drinks often in the past, so I knew that living without them wouldn't bother me.

I grew up with rice and had had a life with it for the last sixty years. I could imagine some hardship if I quit eating rice immediately. I asked myself, "What will be the replacement for rice in my coming days?" However, for the sake of my health, I knew that I had to do whatever was necessary to overcome this stumbling block.

To make my life a little easier, I decided to take an incremental approach in weaning myself from rice. I immediately cut down the daily consumption of rice from two bowls to one. To fill up my stomach, my wife prepared more vegetables, chicken, and fish. Yes, she put slightly more meat on the dining table than she had before. In place of breads for breakfast, my wife cooked eggs, sausages, hams, and cheese.

I also began to cook breakfast on some mornings. I usually scrambled three eggs together for my wife, my younger daughter, and myself. I would make two slices of panned eggs. In between them, I placed shredded cheese and green, red, or yellow peppers to make a breadless sandwich. I would slice it into three pieces for us. Sometimes, I would use spinach instead of peppers. Sometimes I would use turkey or boiled ham in between, to make ham-and-egg sandwiches. Other mornings I would make a whole piece of panned egg. After the eggs were well fried, I would put sausages, peppers, and mushrooms on for an egg-pizza.

By the end of November 2002, I was cooking breakfast more and more often for my family. My wife thought that it was nice to let me have fun doing it. Eventually, she loved to have me cook breakfast every morning, because she enjoyed having the extra time for getting dressed.

For the first few days without eating breads and other flour products, smaller portions of rice at each meal and more cheese, I started to experience a few bouts of soft stool and diarrhea. I had more bowel gas than before. I knew that I would not be able to tolerate milk products, even cheese, for the rest of my life.

I continued the diet with a limited amount of carbohydrates, which were mainly from a bowl of rice a day plus vegetables. By the end of October, the bowel gas, soft stool, and diarrhea started to decrease, and eventually stopped.

Starting on November 10, 2002, I cut back my daily consumption of rice further from one bowl to one-third of a bowl. This amount of rice was very insignificant. Two weeks later, I stopped eating rice altogether. My wife had been having a hard time cooking the rice in an automatic rice cooker. It was very difficult to cook a small amount of rice each time. In the past, I had been the one who ate a

82

measurable amount, so she knew how much rice she had to cook. Now, she declared that she was not going to cook rice any more just for our daughter and herself, because they did not eat rice in the quantity I had eaten it.

At the Thanksgiving dinner, I selected foods that contained no flour, rice, or sugar. I still enjoyed the dinner very much. I continued to eat no rice but more cheeses every day. I also ate cream cheese, which I often squeezed an ounce of it into my mouth, between the meals as a snack. By the end of November 2002, I was constipated often. I began to eat more vegetables with high fibers such as celery, spinach, and Chinese cabbages. In the next few days, I felt better and had no more constipation. In the meantime, I worked much less outside of the house, because of the lower temperature. I had to increase the amount of my indoor exercise to make up for the loss of yard work.

Since the beginning of November 2002, I had been walking twenty-five laps on the stair-trail. I tried to do so as often as four out of every five days. In addition, I did two push-ups after walking every ten laps of the trail. To monitor the progress of my clinical symptoms, without using other medical equipment or tests, I continued to monitor the feeling of heavy pressure or ache in my left upper back. I still noticed pain in my left upper back a few minutes after walking on the trail or getting excited, but I was able to exercise slightly longer, before I felt the pain. Although the pressure and ache in my left upper back were not as severe as before, I felt pressure and ache in my left upper chest. Apparently, I hadn't noticed this before, because my left upper back had bothered me so much.

I checked my blood pressure quite often. Usually, I took

it in the morning after I got up. I also took it before and after walking in the trail, and in the middle of the day. By the end of November 2002, my blood pressure was around 185/60. I had learned more about the fluctuation of the blood pressure from my own experience than I had from taking care of my patients, in the operating room for 25 years. The blood pressure of an alert person varies widely, depending on his psychological and physical states. Also, the level of change in his blood pressure readings is very much related to the pathological state of his cardiovascular system. I decided to keep on observing the relationship of my blood pressure reading, to both the state of my exercise and the level of my stress closely in the future.

WINTER 2002-2003

Begin To Lose Weight

As always, I had weighed myself at least two or three times a day. After I had had very few carbohydrates and no rice for the past four-plus weeks, I noticed the needle of the scale started swinging to left. It registered my weight between 184 and 185 pounds by the end of December 2002. "Yes, I started to lose weight!"

I was very encouraged. I began to eat as little carbohydrates as possible. To satisfy the desire to fill my stomach, I continued to use a one-ounce pack of cream cheese, at least once a day. Sometimes, I even ate two or three packs a day. Like the Thanksgiving dinner, I had a good time at the dinners on the Christmas and New Year's Eves. I had no problems in finding foods I could eat with low carbohydrates. I continued to check my weight several times a day. By the end of February 2003, my fasting weight had come down to around 181 pounds. Great, I was still losing weight! Now, I could notice that my pants were slightly loose. I had been wearing pants with a 36-inch waistline.

Push For My Physical Stamina

Since the New Year, I had begun to walk a little bit more on the stair-trail. I walked 30 laps in each session, and did 2 push-ups after every 10 laps. I still felt heavy pressure and ache in my left upper chest after walking four or five laps. I noticed the pressure and ache were not as severe as before. However, I did not want to give myself a false assurance that could fool me into a big trouble.

On February 8, 2003, my wife and I flew to California to visit our son's family. Our new granddaughter had been born four weeks earlier. During the stay, we helped our son and daughter-in-law take care of our grandchildren. They also took us to several scenic places for sightseeing.

On February 16, they took us to San Francisco. After visiting a park and a museum, we went to Chinatown. It was a chilly day, especially in the late afternoon, when the temperature dipped with drizzle and a cold breeze. I worried that my granddaughter might get cold when we had to run from one building to another on our way from the parking garage to Chinatown. I took my granddaughter inside my slightly oversized coat, and zipped the coat up to the point below my neck, so she would have enough air for breathing. Although I noticed the weight of my granddaughter inside my coat, I ran alongside the others without much discomfort. I could really appreciate the improvement in both my stamina and my symptoms.

Both my son and daughter-in-law are very thoughtful. While we were in California, they accommodated me with the choice of carbohydrate-restricted diets. I also walked up and down the stairs of their home every day for exercise. My wife and I returned home on February 17. Both of us had upper respiratory infection from the contaminated air inside the cabin of the airplanes. Both of us recovered in about one week. As in the past, my wife and I had to take penicillin for a week, after we began to cough up yellowish mucus.

I resumed walking on the stair-trail again as soon as we got home. By the end of February, walking on that stair-trail had become slightly easier. I also noticed that I could walk slightly faster. Thus, each day it took me a few minutes less to finish the exercise. I was confident that I could gradually increase the number of laps in each session. Eventually, I would walk 40 laps each time I exercised.

Walking on the stair-trail really benefited the muscles of my legs and hips. It was an excellent but laboring exercise. However, I knew I must walk carefully to avoid injuries to my ankles, knees, and hips. In the meantime, I needed to increase the exercise for strengthening the muscles of both my upper torso and abdomen. I wanted to increase the number of push-ups and to begin doing some sit-ups. I also began to do flexing my back forward and backward for stretching and loosening my back muscles.

It was very important for me to bend my upper back forward gently with my arms and hands hanging down with no effort. I did not bounce my back in an attempt to reach my toes or the floor with my fingers. Bouncing the back can increase the pressure inside a disc, between two vertebrae and rupture the disc. I also relaxed my back while extending my upper torso and neck backward, with my arms and hands hanging behind my torso. I hoped that when I lost more weight, especially in my abdomen, I would be able to reach the floor with my fingers in bending my back forward.

Blood Pressure Begins To Decline Slowly But Surely

As always, I still weighed myself at least two or three times a day. In the meantime, I also noticed my blood pressure continued to come down gradually during the winter. By the end of the season, my blood pressure was around 175/60 mmHg when I was relaxed. It was 195/68 mmHg, after I had walked for about four or five laps in the trail. At first glance, my blood pressure hadn't seemed that much improved. I welcome the drop of my systolic pressure, although it was only 15 to 20 mmHg lower than that in the previous summer. I was really encouraged.

Food Intolerance Disappears

Since I gave up rice, flour products, potatoes, ice cream, and sugar, and ate more cream cheese and vegetables instead, I had seen some interesting changes in my body. During the last thirty years of my life, I had been intolerant of milk products including my favorites, ice cream and cheese. The intolerance had been worse during the recent decade. After drinking a half glass of milk, eating a small amount of cheese or a food, which had cheese or milk, I would soon experience the buildup of bowel gas, which sent me to the restroom a few times for relief of gas and diarrhea.

After the initial constipation with more bowel gases, I noticed my intolerance was decreasing, and then it disappeared completely. Now, I could eat whatever any cheese and cheese products without a problem. Of course, I did not drink milk because of lactose, which is milk sugar. Did carbohydrate or blood glucose play a role in this change?

Need Less Sleep

The other change that I observed concerned my sleep. I had never slept well. It had been worse since I became a physician who had to take calls day and night. Most of the time, I slept six to seven hours a night. However, I always woke up with tiredness and felt that I needed more sleep, although I could not go back to sleep. For the last couple of months, I had slept for shorter periods than I had before I started the carbohydrate-restricted diet. However, I felt much better and not as though I had lost sleep or needed more sleep. I could continue to work both mentally and physically during the entire day without feeling tired.

SPRING 2003

Increase Physical Activities

When the weather began to get warmer in April 2003, I started to go outside and clean the yard. I raked leaves, picked up tree limbs, and mowed the lawn. I also used a tiller to grade the vegetable garden for my wife. A few wild trees were growing taller than I liked. I cut the smaller trees first, and then I cut a couple of taller ones. They were over 30 feet tall. The job was not easy for me to do by myself, not to mention the risk of damaging the house or other properties. It was very laborious and I knew it should be a good test for my physical state.

I had to look around the tree first to determine how and

where I wanted it to fall down. I tied the tree with one end of a 100-foot rope at the middle of its trunk. I tied the other end of the rope to the tail of my riding mower. Then I drove my riding mower away from the tree, in the direction that I wanted the tree to fall down. I stopped the riding mower when it began to pull the tree, which I noted from the tension on the rope. I went to the opposite side of the tree from the riding mower and started to saw the tree at its bottom slowly. When the tree began to tilt toward the mower, I stopped sawing. I drove the mower away from the tree farther, to renew the tension on the rope. I ran back to the tree and sawed it very slowly. When the tree tilted more toward the riding mower, before it was completely cut off, I moved the mower yet farther away from the tree. I continued to keep a reasonable tension on the rope between the mower and the tree. Finally, the tree fell down to the ground just the way I had wanted it to.

Now, I had to saw the tree into logs and pieces. I hauled away the small branches with leaves to the curb for the city dumpster to pick up. It took me at least four or five hours from bringing a tree down to cleaning up the ground. Most importantly, I felt confident that I could handle more and different physical activities with less pressure and ache in my left upper chest.

Lose More Weight

This amount of yard work really helped me burn my fat and lose weight. I cut three tall trees and a few small ones by the end of May 2003. At that time, my fasting weight dropped to 176 pounds. My goodness, I had lost ten pounds since I began to restrict carbohydrates in my diet. It was really not so bad that I lost 10 pounds in a period of slightly less than eight months.

I continued to notice that my pants were getting too big for me. I had to punch my belts with one extra hole to keep my pants up; the collars of my shirts were also getting looser. For the past decade, I had wanted to dress myself up for some occasions. But I really hated to do so, because I could not stand the choking by the collar of my shirts. I also felt that some of the older jackets were too tight around my chest. Now, I did not mind dressing up, because the collars of my shirts were loose and the jackets were not squeezing my chest any more. What a nice thing, to be able to enjoy this again.

March Toward The Goal Of My Exercise Routine

Since the beginning of March 2003, I began to increase the number of laps on the stair-trail. By the end of March 2003, I had reached 40 laps in each session of exercise. I was now exercising at least four out of every five days. I still noticed discomfort or heaviness in my left upper back after finishing the sixth or seventh lap on the stair-trail. I noticed that the condition of my cardiovascular system had gradually improved. My goal was to be able to finish 40 laps of the stair-trail without discomfort in my left upper back or left upper chest. I had to be patient and take time to shape myself up. Occasionally I still felt some aches.

At times, in the middle of walking and after the onset of discomfort or ache in my left upper back, I would ask myself, "Why do I have to take part in such a silly venture or experiment?" I really wanted to stop doing this exercise and drop the whole project. "Oh, no way," I said to myself, when

I saw the image of myself in a wheelchair being pushed around by my wife. I quickly said to myself, " I have come a long way, and I am beginning to realize some improvement. I am not going to allow a setback."

In the middle of the spring, I added three push-ups after walking every ten laps. After finishing the fourth set of ten laps, I did four sit-ups. For the first few days, the muscles of my abdomen were very sore after starting the sit-ups. Soon I had no problems in doing the sit-ups. By the end of May 2003, I walked on the stair-trail 40 laps in each session. I also did five push-ups before the walk, five more after finishing the first twenty laps. Then I did eight sit-ups after finishing the daily walk.

Exercise And Blood Pressure

It is very interesting to observe my exercise and sweating. When I worked outside in my yard, I noticed changes in my blood pressure, before and after I began to sweat while continuing my work. I took my blood pressure on a Saturday morning in May 2003; it was 178/63 shortly after finishing my breakfast. It is not unusual that the blood pressure is slightly higher after a meal.

Then I went outside and began to bring a tall tree down. Shortly after I threw a rope up to the middle of the tree, I began to feel ache and discomfort in my left upper back. I intentionally worked slowly while carefully monitoring the ache and discomfort to avoid worsening my symptoms. About thirty minutes later, the discomfort had eased, even though I had continued to work at the same pace. About one hour later, I was really hot and I had sweated more. My left upper back felt even better.

About two hours after I started working, I went into the house and took my blood pressure. It was 122/55. Great! My blood pressure had come down! This was because sweating helps carry the heat out that is generated from the work. Sweating is the result of opening up the blood vessels, such as capillaries and small arteries, for releasing the heat by driving out the fluids, especially from the blood. Sweating can also help reduce the blood volume, so bringing down the blood pressure.

What I appreciated the most was that sweating gave the blood vessels a chance to exercise by dilating their lumen. I personally thought that sweating might help soften the vascular wall, especially in the case of arteriosclerosis. For that reason, I always made sure that I sweated while working out.

Steadily Drive Down My Blood Pressure

By the end of May, my blood pressure had come down to around 168/60 when I was relaxed, and around 185/70 when I was excited or after a few minutes into the daily walking. Although the improvement in my blood pressure was slow, it had been progressive! Along with the improvement in my cardiovascular system, I had kept observing any change in my impotence, which had been moderately severe before I started the carbohydrate-restricted diet. Until that time, I also had experienced discomfort in my left upper back during intimacy. During the last three months, I had noticed a great improvement in terms of the impotence. I still experienced discomfort in my left upper back. However, I noticed that the discomfort or ache was only about half as severe as it had been prior to the start of the carbohydrate-restricted diet, or, did not happen as often like before. That meant my blood pressure during intimacy was not going up as much as it had in the past.

Become My Family Morning Chef

It had been about six months since I started cooking breakfast for my family. I had become very good at and I enjoyed doing it. I loved the scrambled egg sandwich without bread every day. Actually, it was an omelet. However, I wanted to try other recipes. Sometimes, I replaced spinach and other kinds of vegetables with mushroom and tomato. They were good, but I did not believe that everyone in my family shared the same taste. Thanks to the changes in my experiments for improving my cardiovascular condition, the switch to the carbohydrate-restricted diet had produced a positive result.

Ideal Weight At The Best Of My Life

Since the beginning of this season, 1 had observed a steady but still slow loss of my body weight. By the end of May, my fasting weight in the morning had come down to 176 pounds. I had lost ten pounds in about eight months. That was slightly more than five percent of my weight last summer, the beginning of my life experiment. I also noticed that my pants were slightly looser; I was wearing 36" pants last summer, which had been tight to my waistline. I wanted to reach and keep my fasting weight at or around 145 pounds; that had been my weight when I was in my late 20's. That might be too thin for me. I decided I would be happy even if I kept my weight at or below 155 pounds.

During my late 20's, I was studying medicine in Japan. My wife and my son joined me there when I started the second trimester of my first year in medical school. My son was 19 months old then. Because there was no other family member with my wife during the day, he became very lonely. He always wanted me to play with him after I returned from school. He even climbed onto my desk and closed my books and asked me to hold him. On an ordinary day, I had to play with him in the late afternoon. Then I bathed him after my wife fed him. I started to study after he went to bed, around nine o'clock in the evening. I finished my study and went to bed around midnight. I got up at five-thirty in the morning and studied another hour and a half. Then I helped my wife get the breakfast ready. I arrived at school by eight.

In the summer vacation of my second medical school year, my elder daughter was born. She brought more fun and made my life even busier. Of course, I have been lucky to have a very capable wife. However, between my studies and family life, I was handling a busy schedule and workload amazingly well. Apparently, my health was at its peak. My slimmer body weight during that time made the most optimal use of my energy. So, I believed that to return my fasting weight, if possible, to the weight that I had during my late 20's would be ideal. Certainly, I understand that my tissues and organs are not going to stay the same all the time for the rest of my life. As said, I just want to keep my body healthy till my last day, like a candle continues to burn quietly until it is exhausted.

Improvement Of My Skin

Other than continuing to use more cheese in my daily diet, my family also ate more vegetables every day. Celery was a particular favorite. I often filled stalks of celery with peanut butter or cream cheese. Or, at times, I mixed the peanut butter and cream cheese.

I have loved nuts for my whole life. I eat boiled nuts, but I could never tolerate dry nuts, especially those fried or roasted nuts and even peanut butter. Every time after eating dry nuts or peanut butter, I used to notice bumps on my skin, like acnes. My face would have more bumps than the rest of my body. They were painful. I could not do anything about them, except apply antibiotic ointment, with or without cortisone (corticosteroid hormone). Each time I had to wait for a bump to mature, with a small yellow pimple in its center. Then I would clean and break the pimple to let the pus out. Then I applied more cortisone ointment to the bumps until the wounds healed. Interestingly, once I started the carbohydrate-restricted diet I could eat all the dry nuts I wanted, without the buildup of these skin bumps. Did carbohydrate or blood glucose play a role in this change?

No Sick Feeling From My Empty Stomach

Although I have never been diabetic, I had felt sick from hunger a few times in my life. When I did not eat my lunch on time, I suddenly became quite weak and sick. That feeling was just like driving a car and watching the needle of the fuel gauge, already on the E mark, continue to move to the left. These are typical symptoms of hypoglycemia (low blood sugar). After eating a regular meal with lots of carbohydrates, especially those with refined carbohydrates, the pancreas responds by producing and releasing insulin. When the pancreas supplies more insulin than the body needs to offset the blood sugar, the body becomes hypoglycemic.

After I had been on the carbohydrate-restricted diet

for at least six months, I didn't have a single episode of hypoglycemia. This was true even after working in my yard for five hours after breakfast without a break. Yes, I did notice my stomach was empty, but never felt sick from it. Apparently the pancreas had learned that I had been feeding myself with much fewer carbohydrates, so the pancreas sent out a much smaller amount of insulin after each meal. In turn, there had been no attacks of hypoglycemia as the result of a flood of insulin by the pancreas.

SUMMER 2003

The First Anniversary Of My Nightmare

As always, I took my blood pressure several times a day. It continued to fall slowly but surely. By the end of the summer, it was usually around 162/58 mmHg when I was relaxed. However, it would go up to 175/68 mmHg after I had walked 15 laps in the trail, or when I was in a rush or excited.

My wife also noticed that I was no longer snoring as I had before I started to lose weight. More importantly, I was no longer having nightmares as the result of airway obstruction, or so-called "sleep apnea". I noticed that I could breathe easier while sleeping on my back! I certainly believe that keeping my airway open during my sleep helped bring down my systolic blood pressure (the upper number of the blood pressure), too.

One-two Punch: Working Very Hard With Very Low Carb Diet

As soon as June 2003 began, the weather was getting

much warmer and, sometimes, it was really hot. I took advantage of the warm weather and continued to work in my yard. Frankly, I worked as much as I could, because I wanted to use body fats as the energy source for my hard work. I needed to catch up on some work that I had put off for the last few years. I targeted a few more trees, which were wild and tall and had grown too quickly. Although I like tall trees around my house, I do not want them growing too close to it. This is especially dangerous during the summer when we are entering the hurricane season. By the end of August, I had brought down four more trees. They were at least 25 feet tall with bushy branches. I had no choice but to work very hard. That required lots of energy. So, I decided to do an experiment.

On June 7, 2003, Saturday, my fasting weight was 173.5 pounds. After I ate my very-low-carbohydrate breakfast, I stripped down to my underpants again for the experiment, as I always did in the morning to get my fasting weight. It was 174.5 pounds. So, I had gained one pound with the breakfast and water.

Then, I worked in the yard, bringing down a tree. I knew I would still feel the pain in my left upper back, if I worked too hard at the beginning. So, I worked very slowly for the first twenty minutes, just like warming up before playing sports or working out.

About 30 minutes after I started to work, I brought down the tree. I sawed it and separated the limbs, and cut the trunk into fire logs. Three hours later, I stopped the work and went into the house. First, I used the restroom. I wiped myself up and removed all but my underpants. When I got on the scale this time, the reading was 169.5 pounds. Wow, I had lost five pounds by just working outside for three hours. Of course, I knew I had lost lots of water while working under the sun. I drank three 8-ounce glasses of water to make up for the acute water loss. I gained back about 1.5 pounds. Then I ate celery with cream cheese and some leftover chicken. I drank a couple more glasses of water, too. After finishing my lunch, I was 173.5 pounds. I had a net loss of 1.0 pounds by working hard in the morning.

That afternoon I hauled away the small branches with leaves to the curb, and I piled up the fire logs. After I cleaned up the yard, I had spent about 8 hours working outside on that day. At this time, I was 168 pounds before I replenished the loss of water. At bedtime my weight was 173 pounds. The next morning, my fasting weight was 172 pounds. I lost 1.5 pounds while working in the yard yesterday, but it was an eight-hour of backbreaking work!

For many weekends during this summer, I had continued to work hard in my yard. All together, I had brought down four tall trees by the end of the season. I had also cleaned the yard and tried very hard to keep the grass alive, although I got an F in maintaining my lawn. Most of the time, the grass grew up in the spring and died before the summer had even begun.

Because my elder daughter was going to marry at the end of August, I expected that many of my relatives would be coming. Some of them would stay at my house. I wanted to fix it up a bit. I peeled the 25-year old wallpaper off the kitchen wall, and painted the wall. It was not very difficult work, but the indoor jobs also helped me continue the weight loss.

Pitch Up My Routine Exercise

During this summer, I continued to walk 40 laps on the stair-trail each day, on four out of every five or five out of every six days. I still noticed discomfort in my left upper back. However, it did not occur until I had finished walking the first 7 or 8 laps. I knew that my cardiovascular system had improved. Depending on how I felt physically during the exercise, I did more push-ups, both before starting the trail and after finishing the first twenty laps. In the meantime, I reduced the number of push-ups at the end of the first and third set of ten laps. I felt that the changes of my workout would be just as good if I could do twelve standard push-ups in one session. There was no reason for me to attempt many more sub-standard push-ups, while struggling with sore shoulders and arms.

By the end of the summer, I eventually did six push-ups before walking in the trail and after finishing the first twenty laps. I also increased the number of sit-ups to 12, after finishing the walk. I did not feel soreness of my abdominal muscles as I had before. With adjustment of the timing for push-ups and sit-ups, I had become much physically stronger. I could finish the 40-lap session without having to stop for a break in the middle of the walk.

Losing Weight Handsomely

By the end of August, my fasting weight was 159 pounds. I had lost 17 pounds during the summer. I had lost a total of 27 pounds since last October. I had reached the goal for weight loss that I had set last year! Now, I could see that the shape of my body had changed from a small barrel into a pear. However, I still noticed the excessive skin fold (dog ears) hanging over the elastic waistline of my pants. I was surprised at that I did not have very much furrowed skin on my abdomen after losing 27 pounds. Besides the "dog ears", the folds between my buttocks and the back of my thighs had increased. I had also noticed that I lost much of my cushion when I sat in a hard chair.

I knew a lot of people who had lost weight and then gained it all back. These people and researchers blamed the ineffective dietary programs for their failure in keeping an ideal weight. I strongly disagree! Those people, who lost weight but gained all the weight back, did not realize the most important reason for them to lose weight in the first place. Losing weight is not a race or a beauty contest, but a serious task in rescuing one's health and life. When I reached the goal of my weight that I set last year, I did not celebrate my success. I understood that I had other health problems and I must continue my efforts to deal with them. I would not have a party to celebrate my small success by eating all the carbohydrate-rich foods I had avoided for the last eleven months. If I had done that, I would have gained back all the poundage I had worked so hard to shake off.

During the month of July, I saw two of my old physician friends at a local COSTCO Club, at different times. One of them used to be obese. The other was just moderately overweight. Both of them had suffered an acute heart attack as a result of coronary heart disease. Both of them had undergone a coronary artery procedure, after which they had both suffered a stroke. When I met them, I was very surprised at how much leaner they were. Both of them told me that they had adopted a low-carbohydrate diet. I was very glad that they did what I had been doing. I also hoped that they would stick to that diet. If they celebrated their success in weight loss, by eating the carbohydrate-rich foods again, they would gain back all the weight.

My Daughter's Wedding

This was one of the most memorable summers of my life. My elder daughter was getting married on August 30 and I had to find a tuxedo for the wedding. I bought a tuxedo in early 1980s, for attending a colleague's wedding anniversary party. Then, of course, I was much slimmer. After the party, I wore the tuxedo on a few occasions. Then I began to get heavier and fatter, year after year until I could no longer wear the tuxedo. In the meantime, my wife kept buying larger and larger dress suits and shirts for me. My wife found my 1980s tuxedo in the closet. It still looked like new, but I wondered if it would fit me. When I slipped my body into the "old" tuxedo, I was shocked. It fit me very well. In fact, it was slight looser than it had ever been. I looked into a mirror, and was so excited about fitting into this tuxedo.

At the wedding, I escorted my wife to her seat. Then, I escorted my daughter to the hand of my son-in-law. I was so excited and moved. I noticed my eyes were moist. After the wedding ceremony, I was busy greeting the wedding guests at the reception before the dinner. I met several of my colleagues who had not seen me for the last six or seven years, since my retirement from the practice of anesthesiology. They told me that they were surprised at my look. They thought I was ill because I had lost so much weight. It was a wonderful wedding dinner. I enjoyed the food, but stayed away from carbohydrates. When I got my plate of wedding cake, I was supposed to eat it for the sake of my happiness, on this important milestone in my daughter's life. Yes, I had to eat the cake, but did not want to disrupt my diet plan. So, I ate a piece of the cake about the size of my thumbnail.

Improvement Of My Airway During Sleep

During the past 11 months, I had become increasingly aware that I had more energy and needed less sleep. I could sleep as little as three or four hours at night, then work the next day until late in the evening without having to take a nap. I believe that the changes are a result of maintaining a clear airway during my sleep.

People who suffer airway obstruction or sleep apnea at night often complain of tiredness during the day, as if they did not sleep much at night. In fact, they may well sleep many more hours than others do. They could be tired because the oxygen concentration of their blood is lower than normal, as a result of t airway obstruction.

FALL 2003

Hurricane Aftermath

This fall was an unforgettable season for me. We were expecting one of the most powerful hurricanes in history on the east coast of the United States. On September 28, 2003, Hurricane Isabel knocked on the door of Virginia. It beat up the trees with gusting wind and heavy rain. It ripped off the power lines and left the people of my area without power for a few hours for some, to days or weeks for others. Our house lost power for nearly eleven days. We were fortunate to have an electrical generator in our house, so were able to restore power for cooking and refrigeration. However, cranking up the generator every few hours was no fun at all. Sometimes the engine started easily while at other times it did not. After cranking the engine so many times, I really noticed discomfort in my left upper back. However, the discomfort was not as severe as that would have been before the start of my exercise and weight loss program.

After the hurricane passed our area, I found only one tree had come down in my yard. Thanks to my hard work during the summer, there was no longer any tall tree standing near my house. On the first weekend after our regular power returned, I cut down the tree and took care of the limbs and leaves. I did not feel much physical stress at all, compared to what I had before.

Redesign My Trail Route

I continued to walk 40-lap sessions on the stair-trail. But, I did not want to walk across the tiled kitchen floor so I decided to change the trail. I would now walk up from the bottom of a staircase to the second floor, across the room over the garage to the main house, and then down the other staircase. Then I would walk back on the same trail. After changing the pattern of the trail, I immediately noticed more of a physical strain when I was walking in the new pattern. That was because I had to walk up stairs immediately after I had walked down them. I felt discomfort in my left upper back a little bit sooner than before. However, while closely monitoring myself, I thought it was a good change for my cardiovascular rehabilitation. So I continued the new pattern, having told myself that I would stop if I could not tolerate it. By the end of the fall, I indeed had tolerated the new pattern well with less discomfort in my left upper back. I started feeling the discomfort only after having walked 7 or 8 laps. In the meantime, I continued to do six push-ups before walking in the trail and again after walking the first 20 laps. I also continued to do 12 sit-ups after walking 40 laps.

I had noticed my arms and legs were thinner, but built up with more strong muscles. I even noticed that I was building up the muscles between my chest and the pubic bone (the musculus rectus abdominis).

Blood Pressure Slowly Coming Down

I still continued to check my blood pressure in the morning immediately after I woke up and before bedtime. I also checked before and after exercise. I took blood pressure in the middle of my exercise when I took a break from walking in the trail for doing push-ups. However, I did not do so regularly. The most interesting was to observe the fluctuation in my blood pressure after finishing exercise. My blood pressure would go up as expected. I focused on the systolic blood pressure, because the systolic blood pressure was the one that could rupture an aneurysm, trigger a heart attack, or cause a stroke. Besides, my diastolic blood pressure rarely exceeded 70 mmHg. Because I used an electronic blood pressure machine, I was able to monitor the change of my blood pressure about every minute for ten minutes after exercise. From the beginning of my experiments, I had always sat down for checking my blood pressure. I placed the blood pressure cuff on my right upper arm routinely. After an initial slightly lower reading, my systolic blood pressure went up. My heart rate was fastest at the initial reading of a lower blood pressure. My heart rate began to slow down when my systolic blood pressure peaked up at the second minute. After that, my blood pressure started to come down. My systolic blood pressure always fell much more than my diastolic blood pressure did. The difference between the number of the systolic blood pressure and that of the diastolic is called pulse pressure. The pulse pressure is very important in keeping the circulation to the peripheral tissue.

The readings of my blood pressure on November 25, 2003 (see Figure 25) were typical for before and after exercise. It was 161/62 before starting the trail. It was 175/65 at the first reading after exercise, 180/68 at the second, 170/65 at the third, and 165/62 at the fourth. By ten minutes after exercise, it was 150/57. *HR: heart rate in all figures.*

Date: 11/25/2003

Time	20:06	20:31	20:32	20:33	20:34	20:35
BP	161/62	175/65	180/68	170/65	165/62	166/68
HR	65	105	104	97	94	87
Time	20:36	20:37	20:38	20:39	20:40	
BP	158/65	160/63	155/60	152/62	150/57	
HR	81	78	80	73	68	

Figure 25. My Blood Pressure on 11/25/2003

By the end of this fall, it was around 155/60, when I was relaxed before exercise. However, it would go up to 180/70 after I had walked 40 laps on the trail. After walking into the 9th or 10th lap of the trail, I usually noticed ache in my left upper back. But, it was not as severe as it had one year ago. I also noticed fewer aches when I was excited.

The First Anniversary Of My Carbohydrate-Restricted Diet

About ten days before Thanksgiving, my wife and

younger daughter went to Taiwan for a visit with my in-laws' family. While they were away, I was very busy every day. I had to continue my medical practice; I had to take care of my dogs and myself. Before my wife left home, she prepared all the foods that were low in carbohydrates for me to eat while she was away. I ate boiled eggs, cheeses, chicken, seafood, fresh sausages, and plenty of vegetables. By the time I left for California to join my wife and daughter at my son's home, my fasting weight had come down to 156 pounds. Wow, I lost 30 pounds in 13 months! Of course, I realized that the weight could come back and could go up and down.

Some of my former patients who came back to me for chronic pain treatment noticed the remarkable changes in my appearance and physical stamina. They wanted to know what had happened to me. I could not remember how many times I had to tell patients about the experience with my medical problems and my decision to treat the problems not with medication but with diet. During the past decades, I had seen more and more patients who were overweight or obese. Many had diabetes mellitus (DM) and cardiovascular problems. A few of them suffered from strokes.

In reviewing research reports, I learned more about hyperglycemia (high blood sugar). Not only can hyperglycemia cause weight gain, obesity, and diabetes. It also can cause inflammation for developing many diseases in the body. Many of these inflammatory reactions cause pain. Those include, but are not limited to, fibromyalgia (fibromyositis or muscular rheumatism in the old terminology) and arthritis.

It was advantageous for at least some of my patients

who were overweight or obese to lose weight with a low-carbohydrate diet. With their enthusiasm, I began to share my personal experience with them. I suggested that they consult with their family physician, internist, endocrinologist or cardiologist before starting the new dietary regimen. Interestingly but not surprisingly, they too lost weight. Those patients who were diabetic and using oral diabetic medications or insulin were able to reduce the dosage of, or discontinue, the medications or insulin. Their blood sugar had been remarkably improved to within the normal range. I really believed that we should prevent or treat diabetes (DM) by first restricting carbohydrates. A carbohydrate-restricted diet could make people feel less hunger and take in fewer calories.

WINTER 2003-2004
Keep The Fire On

After spending a few days around Thanksgiving at my son and daughter-in-law's home in California, my wife, younger daughter, and I came home on December 1. As soon as I got home, I started my regular exercise again. I walked 40 laps on the stair-trail each evening, and tried to walk as often as six evenings out of every seven. I continued to do 6 push-ups before walking on the stair-trail and again after the first 20 laps. I also resumed 12 sit-ups after finishing the stair-trail.

Over the time, the acuteness of the discomfort in my left upper back after starting the daily exercise was abating. The onset of that discomfort had been pushed back slightly. Now, I would notice the discomfort coming on only after walking at least 12 laps on the stair-trail. I believed that this was a good sign of improvement in my cardiovascular condition. It was interesting to observe that the changes on my fasting weight were related to what I ate the day before.

I Love More Food Choices

Because of the food choices in restricting carbohydrates, I became more appreciative of a buffet dinner when I had to eat out. Despite my careful picking and choosing, I could not help but having to eat foods which were mixed with either a small amount of flour or sugar, or both. When I took my loaded weight before bedtime, I would weigh 160 pounds. Then, the following morning, I weighed again for my fasting weight. It was 159 pounds. It was very interesting! I did not lose much weight over night as I always did on any ordinary night. I lost only a pound or a pound and a half after eating the foods outside my home, which were mixed with either flour or sugar, or both.

Keep My Weight Down

On an ordinary day, I would eat about 60 to 70 grams of carbohydrates, which were mainly from the vegetables and fruits high in fiber. My loaded weight would be around 160 pounds before the bedtime. However, my fasting weight on the following morning would more likely be around 156 or 157 pounds. It was incredible to realize a net loss of three or four pounds over night!

Sure, I realize that most of the weight I lost was likely just water. But I did not necessarily have to urinate a lot in the middle of the night. I probably lost water in my breaths during the sleep. But, what had happened when I lost only a pound or a pound and a half of weight overnight, after I had eaten more foods, which were mixed with either flour or sugar, or both? Didn't I expect that I would lose water in my breaths? I did not think that the amount of foods mattered either. Sometimes, I was able to choose more meat, chicken, fish, and vegetables, which were not mixed with flour or sugar, and I would still lose about three pounds overnight. I tried to observe much closely in the future to come up with an answer. During this winter, my fasting weight had gone up and down between 156 pounds and 158 pounds. By the end of the winter, my fasting weight was 157 pounds. It was still 29 pounds less than my weight at the start of the project.

SPRING 2004

Test My Physical Strength

During the last week of February 2004, my granddaughter had become very ill and was hospitalized. At the same time, my son was abroad on a scheduled business trip. My wife and I decided to go to California to help our daughter-in-law. In the wee hours of Saturday, February 28, 2004, my wife and I left for California. During our stay there, I held my granddaughter a lot. With her in my arms, I walked up and down the stairs many times a day. I even did that as my routine exercise; like that I had done at home. Each of the stairs was shorter and slightly deeper than those in my house. In addition, the staircase had more actual steps than those at my house. There was a landing in the middle of the staircase, at the point where the staircase made a 180-degree turn. I continued to walk 40 laps every evening for exercise. In between the laps, I made a short round trip in the balcony overlooking the foyer downstairs. The trip to the balcony gave me a break between each lap, and I did it very well.

On March 2, 2004, we visited a lakefront park. It was a cold afternoon and we were ready to leave shortly after we got there. However, my grandson, who was not quite four years old, wanted to play in a small playground shaped like a boat. Because it was getting colder, I asked my wife and daughter-in-law to take my granddaughter, who was in a stroller, back to the car. A few minutes later, the temperature was dropping very quickly. I was finally able to convince my grandson to leave the playground. As soon as we walked away from the playground, my grandson asked me if I was tired of walking. I said," No, are you?" My grandson quickly answered, " Yes, I am."

I squatted so that my grandson could climb on my back. But he wanted me to carry him in my arms instead. I had to carry him about 1500 feet to the car. I surprised myself that I was able to finish walking without feeling discomfort in my left upper back or chest. I really began to appreciate the hard work that I had done. I said to myself, "From now on, I am going to continue my project to become as healthy as I can be." In the meantime, the visit to the lakefront park in the cold windy evening did not help me well. After sneezing so many times, I felt chills. And my nose started to run with watery mucus.

Catch Upper Respiratory Infection Again?

The next day, after our visit to the lakefront park, my son returned home from his overseas trip. Because of my schedule, my wife and I had to catch a red-eye flight to go home the same evening. The flight was very crowded. My wife and I sat in the last row of the plane. The seat was so uncomfortable because of a thin cushion. I was tired, and fell asleep off and on during the first flight of nearly five hours between California and Texas. Worst of all, the air in the cabin was so bad. I heard coughing from the passengers who were sitting ahead of me. My wife and I returned home in the mid-morning of the next day. It was cold and raining. We could not get into the house, because we had not taken a key to the house. My younger daughter was not home; there we were, outside, wet, and cold.

Keep Up The Heat

I started seeing my patients at noon the day we returned from California. By the time I finished my office schedule, I was tired, sneezing, and blowing my nose. I thought, " I have really gotten a cold. It will be interesting to see how bad it is going to be." Just as before, the course of my cold followed a pattern of sneezing, then nasal discharge with clear mucus followed by a thick one. The thick nasal discharge draining into the back of nose and mouth caused irritation that triggered a cough. As the cough intensified it brought up mucus. In my regular pattern of having a cold up until last year, the entire course would usually take ten days to two weeks. The mucus would become yellowish, and it was necessary for me to take penicillin before the end of it. Surprisingly, I did not cough up as much mucus as I would have in the past. The course of this episode took only about a week, and I did not have to take penicillin this time!

As soon as I got home from the trip, I started exercising again. I could not push myself to do more than I had done last winter. I kept the same routine, and walked on the stair-trail 40 laps each evening. I still tried to walk as often as six days out of every seven. I also continued to do 6 push-ups each before walking on the stair-trail and after the first 20 laps. I also resumed my habit of doing 12 sit-ups after finishing the walk.

Although progress had been slow, the discomfort in my left upper back continued to be less severe. I felt this ache only after having walked many laps. With the same carbohydrate-restricted diet, my fasting weight became steady, between 156 and 157 pounds.

SUMMER 2004

The Second Anniversary Of My Nightmare

As soon as the weather started getting warmer, I cleaned out the small oval swimming pool in my backyard, which is just 16 feet by 32 feet. I had to make the pool ready for my grandson who was coming to stay here for a few weeks. I promised him that I would swim with him whenever he wanted.

I had not used the pool for at least five years partly because I had adopted a less physically active lifestyle since I retired from anesthesiology practice. In addition to this, my busy office schedule made me feel tired, after finishing up the office paperwork at the end of the day I had become less and less inclined to doing anything else before bedtime. Besides, I had lost my stamina gradually over the last few years, up until the time when I found out my problems on June 9, 2002. The last time I had tried the pool, I could only continue to swim one and a half laps. That was not much swimming. I had become very discouraged.

So, I shopped around town and surfed online to find a pump, a filter, and some other parts for replacing the broken ones. I had to redo the piping and to make sure that there was no leak in the filtration system. After sweating through a few days' hard work, I had the pool up and running, as it should.

I got into the pool and started to swim slowly. I was not in any distress but felt very comfortable after swimming the first lap. I began to swim the second lap. No problem! I decided to take time to improve my physical conditions with swimming too. I began to swim daily. It was quite enjoyable.

My daughter-in-law and grandchildren came at the end of June. They stayed with the other grandma for most of the time. My four-year-old grandson, Andre, asked his mother to let him stay with us for a few days. I swam and played different games with him in the pool. Playing games in the pool with such an energetic boy could have worn me out easily if had I not started cardiac rehabilitation two years earlier. Now, I could keep playing with Andre and respond to his demands without a problem. I was very glad that I chose to change my diet and lifestyle. If I had decided to take medications for my blood pressure and heart, I would probably have had much less physical stamina at that time. Besides, I would have had to continue taking medications for the rest of my life!

What An Impressive Improvement

In the meantime, my wife and I were busy with our grandchildren and daughter-in-law. We took them to the aquarium, the museum, the zoo, and to the beach. We had good time with them. We also went to Busch Gardens with

Andre. Because our granddaughter was just 17 months old, her mother suggested that we take only Andre to the amusement park.

Shortly before noon on Wednesday, July 7, 2004, my wife and I took Andre to Busch Gardens. That day was very beautiful. The clear sky let the sun bake the earth with its heating rays. My wife and I had not visited this amusement park since our children went away to college. The park was quite changed from our last visit. We did not remember the names of all the places.

We thought that we could let Andre take some exciting rides. After waiting in a long line, it was our turn to take a ride. But we were told that Andre was too young and too little for that ride. Then we waited in the line for another ride. Again, we were told he was too small. Finally, we settled for the rides for the young children with their companions. Andre enjoyed the rides so much. He wanted to come back to the park the next week for more rides and playing in the small paddle pool. It was about the suppertime, we packed up and headed for home. On the way out of the park, Andre saw a cotton candy shop and wanted to have a stick of cotton candy. I told him that he had to go home for supper, and that I would buy him a stick next week.

On July 14, 2004, we visited Busch Gardens again. We quickly went to the rides for the young children, and let Andre have some fun. After getting enough with the rides, he climbed into the monster house where he could cross back and forth on the rope-bridge. I had to follow him all the time for his safety. Again, it was a good test for my physical condition. I passed it.

After playing in the paddle pool, Andre remembered that

I promised him a stick of cotton candy. So, I took him to the train and rode to the station near the cotton candy shop. As soon as we sat down in the train, he was so tired that he fell asleep beside me. When we had to get off the train, Andre was still soundly asleep. I did not want to wake him up, so I had no choice but to pick him up and let him sleep with his head over my right shoulder. I had to walk a few minutes from the station to the cotton candy shop.

When I walked into the cotton candy shop, Andre was still asleep. I bought a stick of cotton candy, then decided to walk back to the children's playground where my wife and my younger daughter were waiting for us. After I walked for a while, I realized that I had to walk down the hill to the riverbank, cross the bridge over the river, and then walk up another hill to get back to the children's playground. The distance was more than a mile. I did not want to walk back up the hill to catch the train again. So, I kept walking.

By the time I had walked back to my wife, Andre woke up and wanted me to take him to the cotton candy shop. He was surprised, smiling from ear to ear when I gave him the cotton candy. I was smiling too. Not only because Andre was enjoying the cotton candy, but also because I had the stamina to walk down and up the hills for more than a mile while carrying Andre. Most importantly, I did not feel any discomfort in my left upper back. This was the best result of my physical tests for the past two years.

During this summer, I continued to walk on the stair-trail 40 laps every day, and about six out of every seven days. To increase the physical challenge, I decided to drop the break with the 12-sit-ups, after walking the first 20 laps. Instead, I did 12 push-ups before walking on the trail and did 12 sit-ups after walking 40 laps on the trail. Initially I felt a little challenge in trying to complete walking on the trail 40 laps without a break. However, I soon noticed that I had less discomfort in my left upper back when I walked on the stair-trail at a slower pace depending on how I felt at the time. Around the middle of July, I could complete the trail without having to take a break or feeling much discomfort in my left upper back. With the music in my headphones, I could walk on the trail without getting too bored.

With my increased stamina, I was able to work outside more often. I continued to trim the tree branches and bring down trees, which had grown too tall and become a hazard to my house. I cut the limbs to the size of fire logs, then hauled and stacked them up at a corner of my backyard.

In the meantime, I continued to cook breakfast for my family every morning. All of us in the house became very comfortable eating the carbohydrate-restricted diet. However, my wife and my daughter did have some foods with flour and sugar once in a while. I found that we ate a smaller amount of food than we had before we started restricting carbohydrates. I was convinced that taking in more fats and proteins satisfies the stomach much better. In turn, we did not need to eat as much. Besides, the low consumption of carbohydrates requires less secretion of insulin from the pancreas. Then, we had no chance of experiencing hunger as a result of low blood sugar (hypoglycemia).

At the end of this summer, my fasting weight was between 155 and 156 pounds.

FALL 2004

The Second Anniversary Of My Carbohydrate-Restricted Diet

So far I had noticed improvement in my physical tolerance for exercise. I could climb the stairs step-by-step in ten laps without much feeling of discomfort in my left upper back or my chest. At times, I even could continue climbing the stairs up and down for 20 laps without having to take a five-minute break.

Around the middle of October 2004, I decided to give myself a little more challenge. I started to attempt 40 laps in one session of exercise without a break at all. I tried to walk slowly up and down on the stairs. Initially, I still had to take about a 30-second break after walking up to the top before walking down in each lap. About one week later, I just needed a break for about ten seconds in turning around, after walking up the stairs each time. Shortening the time for breaks really shortened the time that I needed for exercise. After continuing this model for four weeks, I became much more comfortable in finishing 40 laps in only 25 minutes.

I tried to walk on the stairs everyday, but at times I took one day off every three to five days. Of course, I still felt somewhat frustrated with exercise, especially when I felt the discomfort in my left upper back, even though it was so slight. During the exercise, at times, I just wanted it to end, right then and there. Most times, though, I was able to remind myself that I must be patient in order to enjoy my health again. I thought about many things while walking on the stairs. I was curious about that how long I could continue to walk on the stairs before my joints would get arthritis, or whether I would get arthritis at all. I also understood that exercise really requires a steadfast determination. Every time I got tired and wanted to quit exercising, I quickly saw the picture of myself sitting in a wheelchair being pushed around by my wife. Then I said to myself, "What would be the fun of my life, if I were disabled?" That picture really woke me up and made me continue walking on the stairs.

Besides walking on the stairs, I also increased daily push-ups and sit-ups to 12 times each. I believed that I could do at least a few more of them. But I decided against that idea. The best result in exercise is to do just enough to avoid hurting oneself. If I tried to push the exercise beyond my comfortable limits, I would probably do more for the number and not for the quality. I would pay less attention to the proper movement or position in doing the push-ups. I might begin to notice sore muscles, aching joints, backache, and et cetera.

Family Inspiration

I switched to restrict carbohydrates in my diet in October 2002. Even my wife, Martha, was skeptical about my idea, but she wanted to help me improve my health. Since I began to use daily exercise to improve my cardiovascular system in the summer of 2002, my wife also tried to exercise too.

She had always worked hard at gardening. Her grandfather had a beautiful garden during his lifetime. Apparently, she was inspired when she was a little child. We bought a small house on a one-third acre lot in Richmond, Virginia, in the summer of 1972. She began to grow vegetables, flowers, and plants in her garden. When I entered private practice two years later we moved to another house with a larger lot. My wife continued to grow vegetables, flowers, and plants. Later, we thought that the lot was too small for her gardening, so we moved to another house with a much larger lot. She was very happy with her garden. Whenever she had time, she worked outside in the yard. She trimmed, plumed, transplanted, dug, and hauled. But I always asked her to avoid squatting for a long period when she was gardening, because squatting could cause problems later on to her knees and ankles, and even to her lower back. I suggested that she sat on a small stool while attending her gardens.

In the late summer of 2004, I suggested that she take a walk on the stairs for a few laps. She tried a few times, but could not continue, because her ankles, feet, or knees hurt. I told her, "You really should not push yourself too hard. You just need to walk on the stairs as many laps as you can do comfortably." But she decided that she could not walk on the stairs for exercise, because that was simply too much for her.

On one evening in the middle of November 2004, I was going to start my daily exercise in the stairwell. I found she was walking up the stairs. After she reached the second floor, she turned around and walked down the stairs. I thought, "She must have forgotten something from downstairs." But she just turned around at the bottom of the stairwell and started to walk up again. I was surprised at what she was doing. I asked, " When did you start walking on the stairs again?" She replied," I have walked on the stairs for a couple of weeks." I asked again, "How many laps do you walk in each exercise?" She said, "Twenty." Out of curiosity, I asked her more, "Why did you decide to start walking on the stairs again? I thought that you could not do it." She said with a big smile, "If you could do it for shaping up your body, why couldn't I do the same?" Wow, I could not believe that, I just couldn't! I thought that she would never try walking on the stairs for exercise after we talked about it a couple of months ago. Since she started walking on the stairs for exercise, I have found that she walks better. She has stopped complaining of pain while walking up and down the stairs.

At the end of this season, my fasting weight stayed between 155 and 156 pounds.

WINTER 2004-2005

Continue To Improve My Health

By early December 2004, I could comfortably walk on the stairs 40 laps daily, for at least 5 days out of every 6 days. In the middle of December, my younger daughter asked me if I could take two steps of the stairs in one. I said, "Why not?" So, I began to walk up every two steps in one. I quickly noticed that I had to lift my foreleg higher to walk up every two steps in one; that made me sweat much sooner in the daily exercise. Besides, climbing two steps in one cut the time walking up the stairs in half. That was an excellent way of finishing my daily ritual. At least, that would shorten my time in counting the steps while walking as I had before.

Initially I had to allow myself more of a break after walking up the stairs each time. I felt aching slightly in my left upper back after walking the first six or seven laps. I even took a five-minute break after walking every ten laps of the stairs by going to work with my computer in my study room. When I resumed walking, I could feel that it was easier to walk the next lap.

To take a break after every ten laps, it took me almost 30

minutes to finish 40 laps in a session each evening. For the sake of improving my cardiovascular system, I continued walking up and down the stairs in the new way. I believed that whatever I had done would pay off one day. Gradually, I could walk up and down the stairs easier every evening. I required less time for break after each lap and each set of ten laps.

By the end of January 2005, I could walk up and down twenty laps of the stairs with only break of 5 minutes. I continued to try slightly harder as long as my heart was working comfortably during the exercise. Within a month, it was much easier for me to walk up and down 40 laps of the stairs, by walking up two steps in one. I only had to take a break for a couple of seconds in turning around at the top and the bottom of the stairs. Wow! I made it.

In the meantime, I continued to monitor the physical tolerance of my heart. I felt less and less aching pain and pressure in my left upper back while walking on the stairs. I also found out that I must avoid taking deep breath in the middle of exercise. Taking deep breaths creates negative pressure inside the chest cavity. That, in turn, brings a more rapid blood return to the heart, and builds up a larger blood volume for the heart to pump out in each heartbeat. The larger blood volume in each heartbeat brings up the blood pressure and, at the same time, slows down the heart rate.

Taking deep breaths seems a good idea in slowing down a racing heart. But, the increase in blood volume for each heartbeat can bring up the pressure in the pulmonary capillaries (the smallest blood vessels circulating the alveoli of the lungs). That disturbs the gas exchanges between the pulmonary capillaries and the alveoli (the room where oxygen and carbon dioxide are exchanged). Consequently, the oxygen content in the blood drops. A few times, I have felt fullness in my chest when I was taking deep breaths. Now I took slow but forceful expiration (breathing out) with a small opening of my lips during exercise, and my chest felt better. I continued doing 12 push-ups and 12 sit-ups. It was getting easier to do them. The muscles of my chest, shoulders, abdomen, and back were gradually but surely increasing in size.

In the winter of 2004, I continued to enjoy cooking breakfast every morning. I was still comfortable with the carbohydrate-restricted diet. Now, I weighed between 154 and 155 pounds in the morning.

Another Example Of Reducing Inflammatory Reaction

On the evening of Tuesday, January 25, 2005, I accidentally flossed the both sides of the back of my teeth too hard. I felt a brief, sharp pain instantly. I ignored the pain and went on brushing my teeth. Early in the next morning, I woke up and felt tightness on the right side of the back of my lower jaw. I touched the area and felt soreness and slight swelling. I also felt a little sore on one of the lower molars when I had bitten down a little too hard. I could feel that the tooth was sensitive. I believed that my tooth got infected from the dental floss I used the night before. I checked the back of my left lower teeth, which were slightly sore, but not bad. I decided to observe the development.

On Thursday I noticed the pain in my right lower tooth was not as bad as it had been the day before. The swelling had also started to ease up. However, I felt a much bigger bump on the left lower jaw. The left lower gum was swollen too, and was worse than the right. I was weighing whether I needed medication to prevent a more serious infection. But I decided against that idea. In the past, swelling and infection of my gums and teeth would not go away until I used antibiotics.

On Friday I noticed my right lower teeth had recovered completely without swelling or pain. The swelling on my left remained, but was slightly less in size. My left lower molar was just slightly sensitive. On Saturday the swelling of the left lower gum had come down much more. It was just like a small bump. My left molar was no longer sensitive. On Sunday I felt no more swelling on my left lower jaw. The inflammation of my gums and teeth were taken care of by my body's defense system. I did not need an antibiotic. What was the role of the carbohydrate-restrict diet in this case? Interesting!

SPRING 2005

Strengthen My Immune System

Looking back on the past year, I realized I had not gotten a cold until the middle of March of this year. Usually, I would have been ill with a runny nose and clear discharge for two or three days before the discharge turned thicker and yellowish. At that time, the nasal discharge drained to the back of my nose and mouth, irritating my throat and upper airway. I would begin to cough with yellowish secretion. Before I recovered from a bout of upper airway infection, I would continue coughing for a few days or longer. I always had to use antibiotics to help me fight the bacterial infection and fever.

On March 17, 2005, I saw a female patient at my office. She had been recovering from an upper airway infection for two weeks. When she visited my office, she was still coughing very hard. I was hoping that I would not get infected. Unfortunately, my fear came true. The next day after seeing the patient I began to sneeze and produce a clear discharge from my left nostril. I decided to observe the course of infection. This time I would try to avoid medications, if possible.

After suffering a runny nose on the left side for the first two days, I noticed the right side had started to run. I was disturbed by the discharge and I noticed soreness from frequent wiping of my nose. However, I had no fever or yellowish nasal discharge.

The fourth day after the onset of my nasal symptoms, I stopped sneezing. My nose began to dry up. Interestingly, the nasal discharge had been clear throughout the whole episode. There were no symptoms of the upper airway inflammation such as sore throat, chest congestion, or a cough with yellowish mucus.

Push Harder In My Exercise

During this spring, I continued to walk on the stair-trail in a 40-lap daily session. I had skipped the daily exercise only twice, then only because I was so busy and had returned home late. I still walked down the stairs step by step, and walked up two steps at a time. I found that exercise was getting easier every day. I also needed little or no break between laps. Sometimes, I tried to speed up my walking pace, but I soon felt that my legs getting stiff and tired.

I thought that getting into the habit of warming up before exercise, then taking time during that exercise were good ideas. I also continued to do 12 push-ups before the walk and 12 sit-ups after the walk. After the daily exercise, I usually took a shower. Looking in the mirror at myself with the buildup of muscles with less fat, I could really appreciate having endured the exercise and diet project. I always believe that only persistence and consistency produce the results. There is no shortcut or overnight success!

While continuing my daily exercise, I felt much less discomfort in my left upper back, after starting the exercise. I could walk many more laps in the exercise before feeling any ache. I had also done more work outside my house when the weather was getting warmer. I could almost do the work without feeling any discomfort at all.

My blood pressure had also continued to come down. Now, it was around 145/58 while resting. It was around 175/70 about five minutes after starting exercise, and at the end of exercise. However, it came down more in the morning after a good rest at night. One time it even registered at 107/55.

During the month of May, the weather had been better. Most of the time, it was breezy and fair. I had to mow and water the lawn regularly and apply fertilizers. For that, I had to use the water from the swimming pool, which needed to be drained and cleaned for the summer. Pulling the hoses and sprinklers around in the yard was not that hard, but not a small job either, and was a good way to use some body fat. Also, I had to clean several areas of the lawn for flowerbeds. I was supposed to cut down two more tall trees before the stormy weather started.

I continued to cook a delicious breakfast for my family every morning. I still ate lots of vegetables, some cheeses, fish, chickens, and pork or beef. At the same time, I consistently but slowly lost weight. By the end of May, my fasting weight had come down to 154-155 pounds.

SUMMER 2005

The Third Anniversary Of My Nightmare

June 9th of this year was the third anniversary of my nightmare. I still remembered how panicked, upset, and helpless I felt three years ago. At that time, I discovered that I had developed serious cardiovascular problems. I remembered that I had to make an important decision. Would I go to my colleagues for help with a growing list of medications? Would I start a personal experiment with changes in my dietary regimen and lifestyle? I had different outcomes to consider. On the one hand, the personal experiment would carry a serious risk of suffering from an acute heart attack or stroke. On the other hand, the personal experiment would give me the chance to improve my health by reforming my body constitution without chemicals. If I succeeded in improving my health with diets, I would be able to show others how to restore their health.

By the summer of 2005 I was much improved in terms of not only my appearance and stamina, of course, but also my cardiovascular system. I continued my carbohydrate-restricted diet every day. I lost weight with the diet. The diet improved my blood pressure. There was little or no pain in my left upper back pain when I was in stressful situations, exercising, or working in my yard. Most importantly, the diet improved my general health. I was very glad that I made a risky but effective decision three years ago. All the hardship and mental anguish that I had experienced in the last three years had paid off.

Enjoy Carbohydrate-Restricted Diet

My family of three still enjoyed the carbohydrate-restricted diet. They had not yet fired me from the post of family morning chef. I loved to try different recipes for our carbohydrate-restricted breakfast. I had kept all the ingredients of my recipes the way they were without additional seasoning. My wife wanted to sprinkle a few grains of salt over her scrambled egg. I had added no salt to my menu, although I did not have to avoid salt since I switched to the carbohydrate-restricted diet. To make everyone in my family happy, I added some sesame seeds and a few shakes of seasoned salt to the scrambled eggs. The taste of the eggs seemed much better to my wife.

I used slightly less cheese in the egg sandwich to see if that would affect the duration between breakfast and the time I felt hungry. After doing that for more than one month, I had not noticed the slightest difference in my feeling of hunger. I ate a few kinds of fruits. I always watched the amount of carbohydrates I took in each time. I avoided taking in too much carbohydrate at any one time or in one meal.

Many people likely forget that many tiny meals may still add up to a lot of food. Although I ate foods, which contained very little carbohydrates, I still had to make sure I ate about 70 grams or less of it a day. Otherwise I knew I would see the needle of my scale moving to the right the next morning.

I had tried to stay away from bananas and apples because of the amount of carbohydrates they contain. I ate small portions of peaches and oranges. I also enjoyed fruits like such as honeydews, cantaloupes, strawberries, and blueberries. I loved persimmons, which are very sweet with a high amount of fruit sugar. Until I restricted carbohydrates in my diet, I used to eat two or three persimmons at a time. Now, I no longer dared to do that because it would shoot up my blood glucose.

I had enjoyed eating all kinds of nuts in between meals. As I mentioned before, I had not been able to eat any kind of nuts in the past, unless they were boiled. Until I adapted myself to the carbohydrate-restricted diet, a handful of roasted nuts could make my skin, especially my face, break out as it had with acne when I was in my late teens. Also, I would notice swelling of my gums in the morning after eating some nuts. I believe the inflammatory effect of blood sugar from eating roasted nuts had caused this.

For the last three years, I had enjoyed more pecans, walnuts, peanuts, and other nuts, with no breakout of my face or swelling of my gums. I really believed that eating too many carbohydrates caused inflammation in my body. This is the same reason why diabetic patients are much more likely to suffer inflammation and infections. Their wounds are much more resistant to treatment, even with antibiotics.

Improvement In My Cardiovascular System

During this summer, I continued my exercise routine, usually after supper. Before I started the exercise, I took my blood pressure. My blood pressure had steadily improved. When I got up in the morning after a nice sleep, my blood pressure was between 110-120/50-62 mm Hg (the millimeter of mercury). Some mornings my systolic blood pressure was even as low as 100 mmHg, with my diastolic blood pressure at 50 mmHg. At the end of the day, my blood pressure was usually between 100-130/ 55-65 mmHg. Sometimes, if I had a busy and stressful day, my blood pressure would go up to 140-150/60-68 mmHg in the late afternoon.

After taking my blood pressure before exercise, I got on the scale to see how much weight I had loaded on during supper. I started with 12 push-ups. I did every push-up as well as the standard. I was more interested in doing a smaller number of good push-ups than lots of poor ones. After finishing push-ups and still on my toes and palms, I moved my feet slightly closer to my hands. I flexed the joints of my toes, ankles, knees, and hips a few times. Then I moved my feet a few inches closer to my hands, and repeated flexing the joints a few more times. I went on to walk 40 laps up and down the stairs. I still climbed up two steps in one and walked down step-by-step.

Now I could walk up and down the stairs much more easily, and noticed hardly any pressure in my left upper back. I usually noticed pressure or ache after walking on the stairs 17 or 18 laps during the routine exercise. I also started to sweat after walking more than 10 laps. As mentioned before, sweating in exercise should help open up the blood vessels. After finishing walking on the stairs, I went on to do 12 sit-ups. By the time I completed the whole exercise routine, I was ready for a shower.

Before the end of the summer, I was noticing no pain to my left upper back, but I was feeling some shortness of breath.

March My Weight Down Toward The Goal

Since the end of spring, I had worked very hard attending to my yard. During the first weekend of June, I decided to burn more fats and bring my weight down from 154-155 pound to the goal of 152 pounds, a weight I would be comfortable with. I ate my usual carbohydrate-restricted breakfast on each of the two mornings that weekend. I was so busy at my office that I ate very little for lunches on both Saturday and Sunday. By the time I finished at my office, I was ready to eat my supper. I ate a lot of food, but few carbohydrates and plenty of water to make sure that I was not getting dehydrated. Guess what - I lost another half pound easily.

In August, I had also to bring down two more trees, both over 30 feet tall. One was leaning sideways, so it was quite dangerous. The other was a pine tree, which regularly dropped more needles than we could handle. The circumference of each trunk was close to 30 inches. Sometime earlier I had removed their smaller limbs and branches. This summer had been a long and hot one, so I had to wait for the temperature to come down, before I would clear them out of my backyard.

By working here and there, my fasting weight had come down to 152-153 pounds in the morning at the end of this season. Would I plan a carbohydrate-rich banquet to celebrate my success? Absolutely not! Celebrating the success with a carbohydrate-rich feast was the reason why many people failed to stay in good shape and healthy.

FALL 2005

The Third Anniversary Of My Carbohydrate-Restricted Diet

October 12, 2005 was the third anniversary of starting my carbohydrate-restricted diet. Looking back, I had made excellent improvements in my health during the last three years. Except for a couple of special circumstances, I had not compromised on my food for my taste buds over my health.

On many occasions, my friends and relatives had asked me to try some foods that contained a lot of carbohydrates or sugar. They would tell me how delicious or rare some foods were. I politely declined their offers. However, a couple of times I made a small exception. On August 30, 2003, I ate a thumbnail-size piece of wedding cake at my elder daughter's wedding banquet. On Thanksgiving 2005, I had a very small piece of the birthday cake at a party for my brother-in-law. If I had given myself an excuse for eating the "sweet' and "delicious" foods offered me, I would have gained back all the weight I had lost so far, maybe even more. Then my blood pressure would have gone up again and my heart pain (angina) would have returned. Sooner or later, I would have had a heart attack or stroke. I might have gotten other diseases such as cancers. If I had been lucky enough to survive, I would have been dependent on drugs or medications (the chemicals with side-effects) for the rest of my life.

"During the past three years, I had reviewed more than 350 research articles and was still counting. Those articles pointed out mistakes in understanding the dangerous role that carbohydrates had play in our lives.

Remarkable Improvement In My Cardiovascular System

So far, I lost weight with carbohydrate-restricted diet; and I started cardiac rehabilitation with slowly increasing the level of exercise. Both weight loss and cardiac rehabilitation had improved my physical conditions very much. I continued to monitor my blood pressure regularly. Especially, I liked to observe the relationship between exercise and blood pressure. The following were examples of my blood pressure and heart rates taken on October 15, 2005:

Date: 10/15/2005 Before a routine exercise at 7:30 PM, and immediately after exercise at 7:53 PM, and consecutively for the next ten minutes.

Time	19:30	19:53	19:54	19:54	19:55	19:55	19:56	19:57	19:57
BP	118/61	166/72	155/66	146/72	143/72	153/71	134/71	137/68	135/70
HR	72	104	96	92	91	92	88	88	88
Time	19:58	19:58	19:59	19:59	20:00	20:01	20:01	20:02	
BP	143/69	136/67	128/61	124/66	133/65	130/67	131/64	117/63	
HR	84	86	85	56	87	84	85	84	

Figure 26. My Blood Pressure and Exercise on 10/15/2005

My heart was now able to recover from an exercising state to normal within 10 minutes. For example, my systolic blood pressure (the reading of the upper part of the blood pressure) now could return to below 140 mmHg in less than five minutes after exercise. This was far different from my condition in June of 2002. Three years ago I never dreamed I could achieve such remarkable improvement. Above all I was impressed that I had made the amazing improvement without taking medications.

On November 30, 2005, I finished my supper and rushed up the stairs to check my loaded weight. I took my blood pressure. Huh, it was 163/78 at 7:16 PM. That was high.

Well, I had really rushed a little bit too much! I started to do my exercise routine with 12 push-ups first. Then put on my headband and headphones. I turned on the RIO with my favorite Bruck Violin Concerto. I started to walk up and down the stairs. In the meantime, I was thinking about my personal matters. After finishing 15 laps, I had already started to sweat. By the time I finished walking 40 laps, I was really wet. Then I did 12 sit-ups. I could hear my heart racing; I took a series of blood pressure readings on the upper part of my right arm for the next ten minutes. Again my systolic blood pressure dropped to under 140 mmHg within three minutes. After fluctuating for a while, my blood pressure continued to drop to as low as 118 mmHg at one point.

Have I Reached My Ideal Weight Yet?

During this fall, I had been busy with many different tasks. When something did not go as I wanted, I found that I was disappointed and frustrated. Sometimes, frustrations made me feel less like eating. I was unintentionally pushing my weight down a couple of more pounds to the mark of 150 pounds during the last two or three days of October. "Wow!" I shouted, "I still can lose more weight." But, I soon found that my celebration was premature, because I ate some more carbohydrates from mango and grapes. My fasting weight quickly went back to around 152 or 153 pounds.

Next came an interesting experience. One day at the beginning of November, my wife bought a one-gallon can of boiled peanuts still in the shell. Because the peanuts were preserved with salt water, they were too salty for us. My wife discarded the salty water and rinsed the peanuts with plenty of tap water. Then, she boiled the peanuts again. The peanuts now were not as salty as before.

I love peanuts so much, especially the boiled ones. I began to eat them at the kitchen table. I kept eating them. By the time when I stopped, I found that I had finished close to half of the can. Guess what had happened. My loaded weight at bedtime that day was 157.5 pounds. I weighed at least four or five pounds more than I had that morning. I thought that I might have taken in too much water. The next morning, my fasting weight was 155 pounds. "My goodness!" I said to myself, "I just cannot believe that I could gain so much weight by eating peanuts." Peanuts are good for our health. Depending on the dryness of the peanut when it is cooked, the amount of carbohydrates for each cup of peanuts can vary a good deal. If the peanut is cooked when it is fresh, it contains more carbohydrates. On the other hand, if the peanut is cooked after it is dried, it contains less carbohydrate. Peanuts are rich in calories. A tablespoonful of peanuts contains 60 nutritional calories or kilocalories. I knew the boiled peanuts that I had eaten were cooked when they were fresh in their shells. They contained more carbohydrates than I had expected.

One thing that I have learned is that I should not be too nonchalant while I am eating low- carbohydrate foods. At times I can forget that as pennies can make a dollar, so do other things add up. I must not treat a serving of a low carbohydrate food lightly, simply because it contains only a couple of grams of carbohydrates. I can easily take in 10 grams of carbohydrates if I finish five servings at a time. The more I ate peanuts, the more I was hungry for them. The carbohydrates from peanuts, which I ate, probably stimulated my pancreas to produce more insulin. I should not ignore even as little as 10 grams of carbohydrates. I began to gain weight after eating more than 60 to 75 grams of carbohydrates a day, even when I stayed active during the day.

Changes In My Physical Condition

I continued to notice change for the better in my cardiovascular system. At the same time, I had also observed progress in my overall physical condition during the past three years. Throughout the first three seasons of this year, I had been active physically. I worked in my yard to mow the grass and remove tree limbs. I cut down a couple more trees that had grown too tall. I was concerned about their proximity to my house and the possible damage to if they were struck down in a storm.

The daily exercise of 40 laps, plus the 12 push-ups and 12 sit-ups, was hard work. Because of this, though, I could really appreciate the build–up of the muscles in my chest, abdomen, arms, and legs. I could see the shape of the muscles when I tightened them. I could see the fat underneath my skin was getting thinner, especially my abdomen. I could comfortably put myself in a pair of pants with a 32-inch waist. I had lost four-plus inches of my waistline since October 2002.

I could take stress easily now. Before my physical conditions started to improve, I would notice tightness and ache in my left upper back even when I was just slightly tense. The case in point was watching the last few minutes of an eBay auction, on an item I wanted, during which I noticed a sharp pain in my left upper back.

Now that I had to deal more with both clinical and administrative work at my office, I was under a lot of stress. But, I did not notice pressure or ache in my left upper back at all. I also could walk a distance, without any feeling of discomfort in my left upper back or my chest any more.

Whenever I recalled my experience with pain during the recent years, I still felt very frightened. I was sitting on a time bomb. If I had not accidentally found out how high my blood pressure was on June 9, 2002, I could have been disabled or dead. Coincidence and some insight had saved my life and health.

WINTER 2005-2006

Sensible Dietary Changes

One evening of December 2005, I entered an online forum for weight-loss, and read the posts. I found a couple of interesting, sensible posts by two women. One woman said she simply gave up her snacks. She did not lose very much of her weight at once. However, she was losing her weight slowly but surely. Most snacks are loaded with carbohydrates. Those, which are easy for our intestines to absorb, change themselves into the simplest form of sugar in our blood. That immediately boosts the level of our blood sugar, and makes us feel charged up and ready to go back to work. Too much sugar, with help from insulin, is usually converted into fat and stored in our fat tissue. To no surprise, this woman was able to lose weight because she gave her body fewer calories and carbohydrates. Her pancreas would put out less insulin. That, in turn, would lower the level of her blood sugar less, so she would not feel hungry or want to have more snacks and sweets.

Another woman said that she lost weight by switching her meals between breakfast and supper. She got up earlier each morning to prepare a dinner for her breakfast. She ate steak, fish or chicken, in addition to her usual dishes for her breakfast. Then, she would eat a simple, breakfast-style meal for her supper. With the more-than-usual amounts of fats, proteins, and calories for her breakfast, she would feel full and definitely need no snacks or soft drinks in between the rest of her meals. By cutting down the intake of additional carbohydrates, she was absolutely right on track for losing weight. Much more time would pass before she would realize the emptiness of her stomach. That is because fat and protein would take more time going through her digestive system. Also, as said above, less sugar in her blood would require less insulin from her pancreas. Then, in turn, she would have no hypoglycemic attack or sick feeling of hunger, before she had eaten her lunch.

With my success in weight loss by restricting the amount of carbohydrates in my diet, especially those with plentiful starch and sugar, I had reduced my cardiovascular risks and symptoms, and my physical fitness had improved. Now, I was a firm believer of sensible dietary changes for a healthy lifestyle.

I counseled many of my patients who had known me since the time when I was much heavier. They wanted to know about "tricks" that I had used to lose weight and lower my cardiovascular risks. I advised them to cut back on the consumption of carbohydrates in their diet. Some of them were very enthusiastic and pledged to change their diet by eating less starchy foods. Some of them immediately proclaimed that this kind of diet would kill them. I am very sorry to say that it is a matter of an individual's choice. Do you want to live to eat or to eat to live?

Continue To Improve My Cardiovascular System

During this (2005-2006) winter, I continued to walk up and down the stairs 40 laps at a moderate speed, so that I would not give my cardiovascular system too much hard work. I had noticed that I could do the exercise much easier with little or much less pressure or pain to my left upper back. Occasionally, however, I felt as if I was out of breath inside my chest. In the regular monitoring on my blood pressure before and after exercise, my heart was able to recover much sooner from the stress as a result of that very exercise. My blood pressure had started to come down to below 140 mmHg in about five minutes after the exercise.

I had also used a stethoscope regularly to check my heart and the carotid arteries on the both sides of my neck. Rarely, when my blood pressure was elevated, I could hear humming sounds in the right carotid artery again. However, I had not heard any murmur or bruit (any other abnormal sound). Besides, I had also noticed less shooting sounds of the blood flow in my ears. This was a good sign that my cardiovascular system was continuing to improve.

Improvement In My Physical Status

In addition to my cardiovascular system, other facets of my physical condition continued to improve. I felt I now had a lower level of inflammation in my body, although I had not had laboratory tests for a while. Since I started to use the carbohydrate-restricted diet, I had experienced no inflammation as I had in the past. When I had a cold now, I could recover in two to four days, with only sneezing and a clear nasal discharge. I had had no sore throat, upper airway infection, or coughs with, or without dirty mucus. I did not have to use antibiotics since Spring 2003. I no longer felt like scratching my skin as I had before. The appearance of my skin was much better. Old blemishes were fading away. There was neither sign or symptom of skin infection nor a need for antibiotic ointment or cream for my skin. I had three lesions from keratosis. They were about 15 years or older. A small one was in my left forearm near the left wrist, another small one was on my right forearm near the elbow. The third one, on the front of my right thigh, was about one inch by 3/8 inch and very hard. Keratosis is a skin condition with growth of keratin, a fibrous protein that makes the skin very hard. The lesion can be pre-cancerous. Luckily, all of my three lesions had a few hairs growing out of them; that likely meant the lesions were not cancerous.

Over the years, the lesions were so itchy that I scratched them very hard and peeled off the keratosis once every few months. They bled. Eventually, the wounds would close up, but soon they would grow back into keratosis again. Interestingly, since I started the carbohydrate-restricted diet, I no longer needed to scratch them. The two small ones became soft and disappeared. The lesions were as soft as the normal skin surrounding them. The one in my right thigh continued to soften. This is a result of improving the collagen fibers by lowering the blood glucose level. The carbohydrate-restricted diet reduced the supply of blood glucose. In turn, the blood glucose level stayed within the normal range, reducing the level of inflammation in the body.

SPRING 2006

I Had Been Closer To The Rank Of Obesity In 2002

In early spring of this year, I bought a new bathroom scale. I weighed 159 pounds. "Don't worry." I said to myself, "I had loaded myself all day with foods." When I stepped on the new scale in the next morning after getting out of bed, I was 157 pounds. "What?" I was shocked, "All in a sudden, I gained two pounds? I could not believe that!"

I took out the old bathroom scale and stepped on it, and it did still read 155 pounds. Then, I stepped back on the new one; I weighed 157 pounds again. Now I realized that the old scale was two pounds off. So, my weight on June 9, 2002 was actually 188 pounds, if I followed the reading of the new scale. That meant that I had been closer to the rank of obesity. I began to wonder how accurate the bathroom scale was. I would have to buy a balance beam scale soon, too. I bet that I would weigh more with the balance beam scale.

Continuous Improvement In My Cardiovascular System

During this spring, I had continued my exercise every day. I seldom skipped the routine. At the same time, I had also continued to weigh myself every day, and monitored my blood pressure before and after exercise. So far my cardiovascular system had continued to improve gradually. My heart was able to recover from stress much more easily than it would have in 2002. Based on the readings in my notes on three different days, my blood pressure was still slightly higher than I liked. However, the systolic blood pressure was coming down to below 140 mmHg, usually within five minutes after each exercise.

Since the summer of 2003, I had observed an interesting link between the amount of work before exercise and my blood pressure at the time of exercise. The more work I did and the longer I worked outside my house, the lower the blood pressure that I had before, during, and after exercise. In this situation, I no longer would experience any pressure or pain in my left upper back during the exercise.

I offer my explanation for the above observation. The work I did before exercise drove out a fair amount of water from my body, particularly from my blood circulation. Thus, my blood volume had shrunk and blood pressure was lowered. With a smaller blood volume and lower blood pressure, the flow pressure onto the walls of the heart and the aorta was so reduced that it caused no pain.

I thought that it would be important to gain a better understanding of how the body retains and removes water inside. Based on my experience, sodium was not the main factor in retaining water inside the body. Actually hyperglycemia is. The description of how sodium helps reabsorb water in the kidneys that we have learned is very likely incorrect! This is why diabetic patients become hypertensive. This was why I did not want to use diuretics to force water out without first keeping my blood glucose within the normal range. Besides, for four months of 2002, restricting fluid and a no-salt diet did not help lower my blood pressure at all. As a matter of fact, I had not had to restrict salt in my diet since I began restricting carbohydrates. However, I did not add salt to my plate after the food was prepared.

SUMMER 2006

The 4ᵗʰ Anniversary Of My Nightmare

Since June 2002, I had continued to follow the carbohydrate-restricted diet, to work very hard at exercise, to observe improvement in my cardiovascular system, and to watch the changes of my weight like a yo-yo. I read as many articles as I could find. I still weighed myself at least two or three times a day. I continually tracked the relationship between the changes of my body weight and the possible factors involved in bringing those changes. It was quite interesting.

For example, I found out that I could lose about 0.25 pound after each session of my regular exercise. I also found out that I could lose 0.25 pound each hour between my supper and the next breakfast, if I did not eat or drink anything after supper. As I noted about last season, when my blood pressure was lower I would feel no pressure or pain to my left upper back during exercise. My blood pressure would be lower during exercise if I worked outside for a few hours shortly before the exercise. I thought that the way the body handles water was very intriguing, as I realized we should not just use diuretic to force the kidneys to remove water. We should first understand how the body retains and removes water. We should also help the body use the natural mean(s) of restoring its water regulation. When the body can reduce the load of water, inside especially in the blood circulation, the blood pressure will come down.

My fasting weight was 155.5 pounds, with the new bathroom scale, at the end of this season.

Watch Out The Dynamic Changes Of Blood Pressure

Blood pressure is very dynamic. It changes very rapidly depending on the state of the body, both mentally and physically. We have to understand the situations that change blood pressure. Thus, we must monitor our blood pressure frequently every day. That can prevent avoidable disasters that can result from dynamic changes in blood pressure.

Over the last four years, I had learned a lot about blood pressure and the body, through the frequent monitoring of my blood pressure. Up until June 9, 2002, I had drunk 12-14 8-ounce glasses of fluids every day. For the next four months, I had limited drinking fluids to 7-8 of 8-ounce glasses a day, while monitoring the color of my urine. I wanted to make sure that I was not becoming dehydrated. Because water restriction and a no-salt diet had not helped me bring down my blood pressure, I stopped my first experiment. The first experiment was based on the theory that we use in the traditional (Western) medicine.

Starting on October 12, 2002, I drank water and used salt without restriction. However, I usually drank more in the late afternoon and early evening. After being absorbed through my bowels, water went into my circulation, expanded my blood volume, and raised my blood pressure. My blood pressure had been much lowered to the normal range while I was relaxed. But, loading my body with more water made my blood pressure higher, at least temporarily, until my body could drive out the excessive amount of water.

Because of loading too much water during the early evening, the blood pressure reading at 8:08 PM on

06/11/2006 was 147/62 mmHg, before I started exercise. After walking the first ten laps, my blood pressure reading was 144/63 at 8:21 PM. After finishing the second ten laps at 8:27 PM, my blood pressure was 174/75. Because the blood pressure was high, I cancelled walking the last twenty laps. My blood pressure did come down to 141/71 at 8:31 PM. But it did not go lower than 140 mmHg for the systolic pressure through the end of the monitoring at 8:36 PM. When I did not overload my body with water, my blood pressure was nicely lowered to the normal range. I also observed improvement in blood pressure, even shortly after excitement. As in my observation, my blood pressure was much improved after working outside my house for a few hours before the exercise. I did not experience ache in my left upper back at all throughout the exercise.

In the continuing blood pressure monitoring, I learned the impact of overloading the body with fluids. Overloading my body with fluids in the past decades was probably one of the causes for my severe systolic hypertension (the upper number the blood pressure reading.) I knew that my hemoglobin (concentration) was also reduced because of a diluted, larger blood volume. Sure, I should have stopped drinking so much fluid (12-14 of 8-ounce glasses) daily long before my problems started. The important point was to understand why I was taking in so much fluid in the first place, other than from heeding the wrong advice of "drinking more fluid is better for you."

Now I believed that hyperglycemia makes us drink more fluid and causes more fluid retention in the body.

FALL 2006

The 4th Anniversary Of My Carbohydrate-Restricted Diet

In late August, one of my friends gave me a 25-pound box of roasted, unshelled peanuts. I already knew that boiled, unshelled peanuts could put weight on me, if I ate more than a handful of them. So I was very mindful that I was not going to make the same mistake. But I loved all kinds of nuts, especially peanuts.

During the extended Labor Day weekend, my family and I were watching movies in our kitchen. Sitting in the kitchen or family room watching a movie always made us want to drink or eat snacks or both. This day was no exception, so we thought about eating the roasted, unshelled peanuts.

I like watermelon seeds very much, because I have to spend more time in cracking them before I can eat them. I would have to spend hours eating them before I could put enough carbohydrates for me to gain weight. Besides, the dry, roasted watermelon seeds are very rich in fiber and low in digestible carbohydrates. They do not quickly bring up my blood sugar level. On the other hand, the dry, roasted peanuts are supposed to be rich in fiber, but also higher in digestible carbohydrates than the watermelon seeds are. Besides, shelling peanuts is much easier than cracking the watermelon seeds.

When we had finished the two-hour long movie on that Friday evening, I realized that I unshelled more than three handfuls of peanuts. In other words, I took in about 20 grams of carbohydrates. Of course, I had swallowed one and a half 8-ounce glasses of water while I enjoyed the peanuts.

The next morning, the scale showed 155 pounds. I gained

one pound on my fasting weight. I did not worry about it, because I usually allowed my body weight to swing within one and a half pounds. I worked very hard in my yard all day Saturday. I weighed myself again after finishing my yard work that afternoon. It was 153 pounds before I drank two 8-ounce glasses of water. Great! I thought that I had shed the extra pound. That evening we watched another movie. Again, we enjoyed the peanuts. I took in another 20 grams of carbohydrates with a couple of 8-ounce glasses of water, too. Before I went to bed, I weighed 159 pounds. I thought I would lose weight overnight by fasting.

The next morning, my fasting weight was 156 pounds. I could not believe it! I thought that I managed to shed the extra pound by working hard all day in the yard the day before. Now I had gained one more pound over yesterday morning.

We had little rainfall for about one month during the summer so I had very little to do in my yard. I just had to pick up dead tree limbs and sweep the driveway and street. Despite my daily exercise with climbing the stairs 40 laps plus push-ups and sit-ups, I had not been able to shed the extra two pounds of weight. It was quite frustrating!

Led by my own frustration, I was more convinced that weighing everyday is very crucial. If someone really wants to lose weight and keep his weight steady, he must weigh himself at least twice every day. By doing so, he will not let himself gain back the lost weight, before he figures out what has gone wrong in his weight loss efforts.

I Am Trying To Lose Weight Again

After gaining two extra pounds during the first week of

September, I decided to be more cautious with my diet. The only difference in my diet during the past two weeks was that I ate more peanuts for snacks. I had already stopped eating them after gaining the extra two pounds. I wanted snacks, not because I was hungry, but because I was not busy at that time. It is important for me to either stay busy between meals or eat snacks with low amount of carbohydrates. One more important reminder: Eating too much of low carbohydrate food still can add up to enough carbohydrates for gaining weight. So, I stayed away from peanuts and snacking for five days. I finally saw my fasting weight slowly coming down, reaching the mark of 154 pounds.

Perfect Wound Healing #1

On November 4, 2006, I was up on an aluminum flexible ladder working on a sign at my office. The ladder had three joints for folding or straightening to adjust its length. To make the ladder a stable base to support my body, I made it into an L shape with two sections of the ladder on the parking lot. Suddenly I noticed the ladder was sliding down on the parking lot with the base slipping away from the sign. I went down with the ladder very quickly. Luckily, I was not thrown off the ladder. Also, I was wearing a pair of jeans at the time. I fell against the edge of a curb and got a 2-inch long gash on the front of each of my lower legs. The wounds were about one-fifth inch deep. They were painful! But, the pair of jeans helped avoid a direct contact between the curb and my legs, or I would have had to worry about tetanus infection. I quickly ran into my office and cleaned the wounds with alcohol swabs. I felt a sharp stinging and burning, but I went back out and continued working on the sign.

That evening I took a shower as usual. I wanted to see how the wounds would progress, so I left the wounds open, without applying any medication to them. Initially the wounds became filled with blood crusts. The blood crusts gradually turned darker day after day. At the same time new skin closed in from the rim of the open wounds. New tissues also grew up from the inside of the wounds. So, the darkening blood crusts were raised up slowly to fill up the opening of the wounds. Three weeks after the accident, the wounds were completely closed without infection. They are as soft and smooth as the normal tissue around them, except that they looked darker with pigment inside the skin.

The most distinct differences in this healing process as compared to that I had experienced in the past were

- very little inflammation
- no infection even without applying antibiotics to the wounds
- no hard scars or excess tissue from the wounds

WINTER 2006-2007

Running Up The Stairs

Since the middle of October 2004, I had walked up and down the stairs, step by step, 40 laps of daily exercise. I had continued to improve my physical strength. In February 2005, I started to walk up two steps in one and walk down slowly 40 laps in an exercise routine. I had also continued doing 12 push-ups and 12 sit-ups. Starting on January 6, 2007, I began to increase the level of exercise by running, instead of walking up the stairs. I would run as fast as I could, two steps in one, and walk down slowly, step by step, for 40 laps of exercise.

Initially I had thought that the challenge would be too hard, although I loved shortening the time I spent on exercise. It wasn't so bad, once I had tried it a few times. It required putting out more energy in running up the stairs and that helped expend more calories. I figured that each exercise session would burn about 450 calories, based on my weight and the type and duration of exercise.

As I kept running the stairs, I could not help but wonder how long I could continue doing that. I would be amazed to be able to continue the same exercise level into my 70's, 80's and beyond. At the time I also wondered whether my knees would give out on me, because of the degree of the exercise. However, I believed that they would probably stay well as long as I did not get an injury. Getting a serious injury was now less likely because my body had a very low level of inflammation (even below the normal range), thanks to my carbohydrate-restricted diet.

Exercise, Vegetables, And Fruits

From time to time, I read articles in the newspapers and journals, including those for medical professionals. Often the articles reported that eating more vegetables, fruits, and doing exercise helped people stay healthy. In my experiments and observation over the past four-plus years, I had gained a better understanding of the real roles of vegetables, fruits, and exercise in our health.

I used to love fruit very much. But, I was no longer eating all kinds of fruit. I ate only the fruits with a low amount of digestible carbohydrates, which are usually fructose or fruit sugar. I also avoided eating a big amount of these. It may sound funny that each time I would eat only a quarter or one half of an orange and a few grapes, but no bananas.

Fibers are carbohydrate, too. However, our small intestine, in general, cannot easily digest and absorb fibers, and lets the fibers go as the main part of the waste. Besides, eating more vegetables rich in fibers increases the volume of the meal, and helps reduce the amount of other types of foods consumed. I ate more vegetables with high amount of fibers too. My family of three was eating celery and other green vegetables with leaves and stems daily.

Eating too much cheese and fat had formerly caused constipation for me. At this point, I was dealing with no constipation, despite the fact that I had been eating a fair amount of different cheeses every day.

Perfect Wound Healing #2

Three months after the accident on November 4, 2006, in which I wounded the fronts of my lower legs, I still had seen no signs of infection or overgrowth of the scars. They remained smooth and without inflammation. The only sign that reminded me of the injuries was the darker skin of the old wounds. In fact, the color of the skin of the wounds was getting lighter all the time. The entire healing process indicated that my body had developed a strong ability to defend itself from infection. My body also had a gentle inflammatory reaction in case of injury and infection. This observation underlines the findings in many clinical studies that a carbohydrate-restricted diet can build up the immune system and healing functions.

SPRING 2007

Body Mass Index vs. Height Waist Index

Before I started my diet for losing weight, I had a tummy that stuck out slightly. I wore 36-inch pants that felt a little bit tight. When I lay down on my back my tummy was flat like a runway connected between the end of my breastbone and pubic bone. Now I had weighed 155-156 pounds for quite some time. I wore 32-inch pants that are still a little bit of loose. When I lay down on my back, my tummy was concave, like a recess in the earth, from the end of my breastbone to my pubic bone. The thickness of the fat underneath the skin of my tummy was about one centimeter. However, based on my current weight, my Body Mss Index was still about 23.5. At least I had pulled away from the territory of "overweight." To address the risks of cardiovascular problems, cancer, and other serious diseases, we should develop an index based on the size of the waistline. I would start to develop a Height Waist Index.

Continue To Improve My Health

As of spring of 2007, I had followed the carbohydrate-restricted diet for four and a half years. I noticed a continuing improvement in my health, especially in my immune system. During these few years I have had no infections that might have required antibiotics. I flew to Taiwan between the end of March and early April. I did not contract any infection in my upper respiratory tract, as I had in the past after breathing recycled airplane compartment air.

Some of our white blood cells are responsible for defending our body from the invasion by foreign organisms such as bacteria and viruses. They are more capable of fighting the invading organism when our blood glucose level is between 50 and 100 mg%. Only when we restrict carbohydrates in our diet can we keep our blood glucose level within the most ideal range.

In this season, I continued to run up 40 laps of the stairs, two steps in one, and to walk down them, step by step. I also continued 12 push-ups and 12 sit-ups everyday, except for the time when my wife and I visited Taiwan. When we were in Taiwan, we walked a lot. That was enough for my routine exercise. The difference was that I had felt no discomfort in my left upper back or my left upper chest in walking. However, I still felt mild pressure in my left upper back occasionally after running 20 laps of the stairs. When I began to sweat after about 25 laps, that pressure started to ease off.

My blood pressure had been as steady as it had been in the last season. Most of the time it was running at about 120-130/50-65 mmHg when I was relaxed. It went up to 145-160/60-70 mmHg when I was excited. During the month of May, I worked outside my house quite a bit. My blood pressure did not go up as much when I exercised shortly after working outside, nor did I feel any pressure or discomfort in my left upper back. This observation underlined the importance of the level of blood pressure for the patient with coronary heart disease.

My skin continued to improve. The two small keratosis lesions, one on each arm, completely disappeared. The big one on the front of my right thigh was getting softer, only an area of less than one quarter of an inch was slightly harder than the rest. All the old scars were getting smooth and their color was fading. When I accidentally got a new cut, I did not see infection or inflammation with it. I seemed having had a nice season with better health.

SUMMER 2007

The 5ᵗʰ Anniversary Of My Nightmare

So far my blood pressure had been within the normal range most of the time. Occasionally I still experienced very mild pressure or pain in my left upper back, in the middle of running the stairs. That would most often occur after I had loaded my body with too much water in the late afternoon. I knew I had to maintain a low body weight in order to keep my blood pressure within the normal range. Nevertheless, I was very grateful for what I had accomplished already in terms of changing my diet and health.

Do I Gain Weight Again?

After keeping my fasting weight between 155 pounds and 156.5 pounds, I was a little bit tired of watching the scale. I was still using a spring-operated bathroom scale. I didn't feel I could trust its accuracy. I saw the reading would go up whenever I bounced on it a few times. The more I bounced on it, the heavier the weight reading was.

On the evenings of July 27, 28, and 29, I sat in the kitchen watching movies with my wife and daughter. I ate more than a handful of roasted peanuts each evening while watching movies. This was a bad habit that I had always wanted to break. Not surprisingly, I had gained one-and-a-half pounds on each of the next three mornings. Eventually, I weighed 159.5 pounds according to the bathroom scale, and I could not manage to lose a bit of it. Oh, NO! I tried to lose weight, but could not. The weather was so hot and I sweated a lot. I thought that I should lose more water than usual and that in doing I would lose weight.

Because I had doubted the accuracy of the bathroom scale for quite a while, I bought an electronic bathroom scale. I could not believe that the electronic scale gave me exactly the same reading, even after I emptied my bladder and bowel. In my opinion, this electronic scale was even worse than the scale with the springs inside. So, I ordered a balance beam scale and got it on Wednesday, August 8.

The balance beam scale gave me a bad news about my weight. It showed that I actually was at least two and a half pounds heavier than the reading from the present bathroom scale. My loaded weight was 161.75 pounds on August 8, not 159 pounds with the current bathroom scale. As compared to the first bathroom scale I used on June 9, 2002, there would be four-and-three-quarter-pound difference. My goodness! This meant I weighed 190.75 pounds on the day of my nightmare, June 9, 2002. My Body Mass Index should be between 29 and 30, not low 28, as I had originally thought.

Based on the Height Waist Index that I was still developing, I got 25.88. This meant I was halfway to obesity five years ago. While trying, in my frustration, to find an excuse for the weight gain, I quickly realized that since the middle of July I had sat a lot. I had picked up the momentum for writing again and I had started to spend more time on this book. Also, I spent very little time working in the yard, partly because of the hot weather. The temperature was at more than 90 degrees Fahrenheit every day; on some days, it even reached 100 degrees Fahrenheit. At the same time, I continued to eat the same menu of foods every day. I knew I must do something to bring my weight down.

Bring My Weight Down!

Since I had spent less time physically working outside in the backyard, I needed to reduce the total amount of foods that I ate daily. Okay, I decided that I was going to do just that, to see if I could bring my weight down to the goal, 152 pounds with the balance beam scale.

Starting Friday, August 10, I began to use a half tomato instead of a whole one in cooking for our usual breakfast. I also reduced the amount of shredded cheese from one and a half cups to one cup for the three of us. My wife suggested that she would like to have one link, not two, of sausage for breakfast. So, I also did the same for myself, down from two links in the past to one small link of fresh sausage for breakfast.

For lunch, I still used the usual recipes. I mixed a can of chicken, salmon or tuna, with herbal seasoning, two ounces of cream cheese, and mayonnaise. My wife, my daughter, and I spread the mixture on the leaves of romaine lettuce or stalks of celery. Instead of using 8 to 10 leaves of lettuce, I reduced it to 7-8 leaves.

For supper, I just ate a smaller amount of vegetables. I also stopped eating peanuts after supper. I was gong to see how these changes would help. Interestingly, my weight started to come down bit by bit. On each of the first three mornings, as compared to the morning before, my weight dropped slightly, probably a quarter of a pound each morning. By Monday, August 13, my weight was 160.5 pounds.

I went out to mow the lawn and then cleaned up the yard and driveway under the baking sun. The work helped me lose both water and fat. In the meantime, I continued to exercise daily by running up and down the stairs, plus push-ups and sit-ups.

My weight continued to drop. On the morning of Friday, August 17, I weighed 159 pounds. The next day, Saturday, I weighed 158.25 pounds. On Sunday morning, I weighed 157.25 pounds. On Monday, August 20, I weighed 156.75 pounds. What an exciting result! I was going to try to drop my weight to 150-152 pounds if at all possible, using the new balance beam scale. I did know my blood pressure and physical status would be better at that weight.

Just as I had experienced in past years, my weight continued to go up and down so much that I could not confidently control it. My fasting weight had swung between 158.25 and 155.75 pounds with the balance beam scale. On this weight roller coaster for more than two weeks, I reminded myself (again) that water droplets eventually build a lake, a river, and an ocean. I should never eat anything while watching a movie.

For a week and a half, my stomach felt empty after breakfast much sooner than before. Obviously, taking away one sausage for my breakfast had made the difference. However, I did not feel hungry or sick.

My Exercise

During the hot summer, when the temperature was over 90 degrees Fahrenheit, I did not enjoy working outside my house. It was not only because the temperature was so high, but also because the humidity was often very high.

Mosquitoes and other bugs loved this kind of weather very much.

To make sure that I continued to shape up my body, I exercised every day with little exception. I did not allow myself an easy excuse. I still ran up and down the stairs 40 laps, and did 12 push-ups and 12 sit-ups for my daily exercise. Running up the stairs felt much easier. Since the end of July, I had resumed at least three repetitions of flexing and extending exercises for my back.

When my weight went up four pounds or more from my usual weight, I once again noticed tightness in my left upper back after running 15 to 17 laps. This interesting experience showed that physical comfort level and blood pressure were closely linked to body weight and water overload.

My Blood Pressure Goes Up And Down

So when my body weight went up just a few pounds I experienced a feeling of tightness in my left upper back again. On three occasions, I interrupted my running on the stairs to take my blood pressure immediately. It was 160/65 mm Hg or higher at each of these readings.

In the past, I had noticed that I could run on the stairs without any discomfort in my left upper back or chest. Now that happened only after finishing physical work either inside or outside the house. The positive changes in my cardiovascular state during exercise were related to my physical work before exercise. The changes might have occurred because my body lost water while working. This lowered my body weight, and likely my blood volume, because of water loss.

So, how the body handles its water should be one of the

most important health concerns. For example, a high blood glucose level could make the kidneys take back water and sodium salts. In turn, the blood volume becomes increased; the blood pressure goes up as it does in cases of diabetes (DM).

Another Episode Of Inflammation Caused By Infection

My body had done well at building a strong resistance to infections, since I switched to carbohydrate-restrictive diet. I could cut my skin for small operations with needles, then merely wiped the skin and flesh with alcohol. Sometime I used a small bandage to cover the cuts. Sometimes I did not use any at all. The cuts healed very well without infection.

On Sunday evening, July 29, I noticed a pain in the middle of my left eyebrow. I said to myself that I had probably bragged too much and too soon about how wonderfully my skin could heal. I realized that it was probably some sort of skin infection, such as a pimple. I decided to do nothing with it and not even try an alcohol swab used for pimples.

The next morning, I felt the tip of my nose was very painful. There was a red spot at the tip of it, like a reindeer's nose. "Wait!" I said to myself, "It is not Christmas yet. It is still very hot outside!" Again, I decided to bear with the pain and soreness to see how they would turn out. For the next two days, irritation from both sores continued to decrease. The pimple at the tip of my nose was still quite painful. By Wednesday, the pimple in the left eyebrow was very much tamed and not hurting. The one at the tip of my nose was not as painful and not nearly as red as it had been. By Friday, the pimple in the eyebrow had completely disappeared. The pimple at the tip of my nose was just about to fade away. I seemed that my body was in better shape for handling infection than ever before. In the past, I would have been very miserable with big, painful bumps for a least a week. Then I would have had to break open the wound, squeeze the pus out, and apply some antibiotic ointment.

My First Blood Glucose Experiment

I had blood tests three times since I started my diet program in early October 2002. The test results showed that my C-reactive protein (a protein produced by the liver reflecting the level of inflammation of the body) went down significantly to below 1.0 (Normal Value: 1.0-5.0). My triglycerides went down too. My fasting blood sugar was between 90-100 mg% ("mg%" means a number of milligrams per 100 cc of blood or serum).

I believed that the fasting blood sugar was not accurate, because the samples were taken after I had been up and working. My adrenal glands as well as the α (alpha)-cell of my pancreas must have responded to the level of my physical activities. Adrenaline, from the adrenal glands, and glucagon, from the alpha cells of the pancreas, raise my blood sugar. Fasting blood sugar could not tell how my body continued to act based on my daylong activities including the stress level and mood. So I had always wanted to study the blood glucose level with my body, as part of my continuing research projects.

Thanks to the Bayer Healthcare Company, I met its local representative at my office on August 24, 2007. We had a wonderful discussion on diabetes (DM). She gave me a kit of Ascensia Blood Glucose Monitoring System or Contour. She advised me as to how I could use it in my research. It was very easy to use.

On August 26, I woke up around 6:30 AM and decided to do research that day. At 7:20 AM, my fasting blood glucose was 88 mg%. Later, I had an "ED (erectile dysfunction) evaluation" for the experiment. At 7:44 AM, my fasting blood glucose went up to 111 mg%. This probably was due to stress hyperglycemia, since I had not yet had breakfast. There was no supply of glucose from food. And the scale of hyperglycemia was so small! (Please see Figure 27.)

At 8:10 AM, I started to prepare breakfast, which was a regular one. I had 3/8 of a medium avocado, a small link of fresh sausage, 1¼ scrambled egg, 4/11 of a small fresh tomato, 4/11 cup of shredded whole-milk cheese, 1¼ tablespoonful of soybean oil, and 12 oz of water. Also, I had a multiple vitamin tablet, a 500 mg Vitamin C tablet, and a fish oil capsule.

Before I ate breakfast, my fasting blood glucose came down to 72 mg% at 8:52 AM. I began to monitor my blood glucose levels after meal (postprandial blood glucose or non-fasting blood glucose).

Date: 08/26/2007

Time	7:20	7:44	8:52	9:17	9:32	9:53	10:12	10:32	10:52
BG	88	111	72	82	109	96	88	90	79
Note	FBS	AED	BBKF	ABKF	ABM	Ice W	Cut B	Cut B	-

BG: Blood Glucose mg%. AED: After ED Evaluation. BBKF: Before Breakfast. ABKF: After Breakfast. ABM: After Restroom Break. Ice W: A cup of Ice Water. Cut B: Cutting Box.

Figure 27. My First Series of Blood Glucose Tests on 08/26/2007

Because I did not ask for extra test strips, I could only do six more tests. At 9:17 AM, my blood glucose was 82 mg%,

After that, my blood sugar was 109 mg% at 9:32 AM, 96 mg% at 9:53 AM, 88 mg% at 10:12 AM, 90 mg% at 10:32 AM, and 79 mg% at 10:52 AM, respectively.

From the two-hour monitoring of my blood glucose, I also confirmed that carbohydrate-restricted diet helped me keep my blood glucose level low. That probably helps me stay in good health, including lowering my blood pressure, maintaining my ideal weight, lowering my blood lipids, and keeping my immune system active.

I believed that people who follow a low-carbohydrate diet or carbohydrate-restricted diet have a much smaller increase in blood glucose in response to stress. This should protect them from having acute events of inflammation, heart attack, and stroke.

FALL 2007

Fruits Can Easily Boost Up My Blood Glucose

In early October, Abbott Diabetes Care, Inc. provided me with a Precision Xtra, for blood glucose and ketone testing, with a monitor and ten test strips. Then a local Abbott representative came to my office with more test strips. Now, I could look into how my blood glucose would behave after eating different foods.

My first continuing blood glucose test, on August 26, 2007, showed that I had a normal fasting blood glucose level. In addition, my blood glucose was far below 150 mg% after eating a meal.

I used to eat lots of fruits, especially the sweet ones. One of them was the persimmon, which grew in my backyard. I could eat two or three at one time, before I switched to carbohydrate-restricted diet on October 12, 2002. I wanted to see if they would affect my blood glucose level.

In the morning of November 25, I had a pair of normal fasting blood glucose readings. (Please see Figure 28.) One was 96 mg% at 7:51 AM when I got up, and the other was 97 mg% at 9:03 AM, before I ate my regular breakfast. I finished my breakfast in 5 minutes. After meal, my blood glucose was 103 mg% at 9:18 AM and 97 mg% at 9:34 AM. These indicated that my breakfast had very little carbohydrates to raise my blood glucose level. Then, I ate one half of a medium persimmon (about 55 grams) at 9:35 AM. My blood glucose quickly rose to 123 mg% at 9:50 AM, and 133 mg% at 10:04 AM. It came down to 114 mg% at 10:19 AM and 10:34 AM. There was a low reading of 78 mg% at 11:06 AM, because of a wet fingertip by the rubbing alcohol. The readings at both 10:49 am and 11:08 AM were normal.

Before eating my lunch, my blood glucose was 102 mg% at 1:48 PM. I had a regular lunch with 7 leaves of romaine lettuce and a mixture of chicken, cream cheese, and salad dressing. After lunch, the readings were 107 mg% and 103 mg% at 2:05 PM and 2:18 PM. At 2:19 PM, I ate the second half of persimmon (about 55 grams). My blood glucose went up slowly; the readings were 104 mg% at 2:34 PM, 117 mg% at 2:50, 125 mg% at 3:02, 125 mg% at 3:18, 132 mg% at 3:33, and 138 mg% at 3:49, respectively. It came down to 113 mg% at 4:03 PM, 109 mg% at 4:19, and 112 mg% at 4:31, respectively. I had lettuce for lunch, which might have boosted my blood glucose slightly. However, I believe that eating plenty of fruits is not necessarily a healthy dietary style. This is especially true if I want to keep my blood glucose level below 150 mg% all the time.

Date: 11/25/2007

Time	7:51	9:03	9:18	9:34	9:50	10:04	10:19	10:34	10:49
BG	96	97	103	97	123	133	114	114	98
Note	FBS	FBS			½ PSM	55 gm			
Time	11:06	11:08	13:48	14:03	14:05	14:17	14:18	14:34	
BG	78	105	102	77	107	90	103	103	
Note	FGR W		(L) 7	Rom		FGR W	½ PSM	55 gm	
Time	14:50	15:02	15:18	15:33	15:49	16:03	16:19	16:31	
BG	117	125	125	132	138	113	109	112	

FBS: Fasting Blood Glucose. ½ PSM (1/2 of medium persimmon, 55 gm) each were taken at 9:35 AM and 2:19 PM. FGR w: Low reading of Fasting blood glucose because of wet finger with alcohol. (L) 7 ROM: Lunch with 7 leaves of romaine lettuce, chicken mixture.

Figure 28. Fruit and My Blood Glucose Tests on 11/25/2007

To continue monitoring my blond glucose, I took a pair of fasting blood glucose readings on November 30, 2005. These were normal, 99 mg% at 7.12 AM and 103 mg% at 7:18 AM, respectively.

The 5th Anniversary Of My Carbohydrate-Restricted Diet

My family and I had been on the carbohydrate-restricted diet for the past five years. It was a long time. However, none of us disapproved of the diet. My wife was skeptical about the diet five years ago. Now, she believed in it. Although she would still, but rarely, eat a very small piece of cake or other sweets made of flour, she did not want to eat more of them. She became very conscious of the amount of carbohydrates in foods.

I was still very comfortable with the diet, and I enjoyed it. I had continued to prepare breakfast for my family since October 2002. For the past two-plus years, I had cooked the same recipe every morning, mentioned earlier. Since the end of November, I had added one and a third tablespoonful of cooking oil to the eggs before scrambling. I used the kind of oil we used at that time, such as peanut oil or soy oil. This offered us an opportunity of "eating" some vegetable oil.

The breakfast kept us from feeling hungry for many hours. Sometime, I had to work at my office into the middle of afternoon, and did not have the time for lunch; but I did not feel hungry at all. Our lunch was very simple. We used either chicken or tuna, or, salmon, and mixed them with herbal seasoning. We would add one and a half to two ounces of cream cheese and an amount of mayonnaise to mix the whole ingredients for easy spreading. We spread the mix on the fresh lettuce or celery for lunch.

My wife prepared the supper, which was usually a dish of vegetables and a dish of other foods such as eggs, chicken, fish, pork, or beef. However, we did not have meat or poultry at the dinner table very often. We just took away most of carbohydrates from the meal. In place, we ate more vegetables with lots of fiber. We had been eating smaller meals since switching to the carbohydrate-restricted diet, in comparison to those we had five years ago. However, we were rarely hungry, even if we had to skip a meal.

At the end of this season, my fasting weight stayed around 155.25 pounds.

WINTER 2007-2008

More Experiments On Blood Glucose

After monitoring my blood glucose since August last year, I continued to explore the relationship between the level of my blood glucose and the type of foods I ate. On both December 26, 2007 and February 9, 2008, my non-fasting blood glucose was below 110 mg% in the afternoon and evening. (Please see Figure 29.)

Date: 12/26/2007

Time	14:33	19:28
BG	109	103
Note		

Date: 02/09/2008

Time	15:10	15:26	15:55
BG	83	91	93
Note			

Figure 29. My Blood Glucose Tests Between Meals on
12/26/2007 and 02/09/2008.

On February 24, 2008, I took blood glucose tests for the link between blood glucose level and exercise. I was also interested in finding out if a state with both fasting water load (no water re-supply since bedtime) and fasting blood glucose in the morning would help reduce the magnitude of stress hypertension in exercise, and the elevation of blood glucose.

After getting out of bed at 7:30 AM, I let my dog out and went back to bed until 8:00 AM. I weighed 159 pounds. My fasting blood glucose was 95 mg% at 8:03 AM. My blood pressure from my right arm at 8:06 AM was 128/57, with pulse rate at 90 per minute. (Please see Figure 30.)

Date: 02/24/2008 Fasting Body Weight: 159 pound

Time	8:03	8:22	8:29	8:35	8:42	9:27	9:42	9:57	10:12	10:27
BG	95	88	118	116	115	115	153	114	145	119
Note	12 PUP	10 laps	10 laps	10 laps	10 laps	FBS		TV		
BP	L128/57	R135/55	R138/50	R147/53	R148/60	PC				
HR	90	116	121	118	122					
Time	10:42	10:58	11:13	11:27	11:42	11:57	12:12	12:27	12:42	
BG	113	117	120	118	110	112	112	95	99	
Note			Post-C		Pre-E					

12 pup: starting 12 push-ups at 8:15 AM. 10 laps: Walking 10 laps of stairs. FBS: Fasting Blood Sugar. PC: Finishing cooking. TV: Watching TV. Post-C: After working with a computer. Pre-E: Before eating 28 grams of ground peanuts. BP: Blood Pressure. HR: Heart Rate.

Figure 30. My Blood Pressure and Blood Glucose Levels in Exercise on 02/24/2008 (R: right upper arm. L: left upper arm.)

At 8:15 AM, I started 12 push-ups. At 8:22, I finished the first ten-lap of running on the stairs; my blood glucose was 88 mg%; my blood pressure from my right arm was 135/55 mmHg with a pulse at 116 per minute. At 8:26, I finished the second ten-lap of running on the stairs; my blood glucose was 118; my blood pressure from my right arm was 147/53 with a pulse at 118 per minute. At 8:32, I finished the third ten-lap of running on the stairs; my blood glucose was 116; my blood pressure from my right arm was 138/50 with a pulse at 121. At 8:40, I finished the final ten-lap of running on the stairs; my blood glucose was 115, my blood pressure from my right arm was 148/60 with a pulse at 122 per minute. After doing 12 sit-ups and bending my back forward and backward, I started to prepare breakfast at 8:50. At 9:27 AM, I was ready for breakfast; my fasting blood glucose was 115 mg%. The rise of my blood glucose was probably because of stress hyperglycemia. I ate 1-1/3 scrambled egg with 1/3 cup of shredded cheese, a sixth of medium tomato, 1-1/4 tablespoonful of peanut oil, one third of an avocado, and a small link of fresh sausage, plus 7 ounces of water. I finished my breakfast in 5 minutes. My blood glucose went up to 153 mg% at 9:42. Then, it fluctuated slightly with readings of 114 mg% at 9:57, 145 mg% at 10:12, 119 mg% at 10:27, 113 mg% at 10:42, 117 mg% at 10:57, 120 mg% at 11:13, and 118 mg% at 11:27. I ate 28 grams of grounded roasted peanuts (160 calories), plus one and a half cups of light tea (10.5 ounces.) My blood glucose was 110 mg% at 11:42. Following that, my blood glucose was 112 mg% at 11:57 AM, 95 mg% at 12:12 PM, and 99 mg% at 12:42 PM.

During the entire course of exercise, I had no feeling of tightness or pain in my left upper back or chest; my blood pressure did not rise as much as it had in the past either from my regular evening exercise. At the same time, I did not see stress hyperglycemia until briefly after finishing my breakfast. I thought, "I shall be in very good health if I continue the carbohydrate-restricted diet and avoid overloading myself with water." As I experienced in the past, too much water was dangerous to my health!

The Case Of Stress Hyperglycemia

Stress hyperglycemia is dangerous. It is a state of high blood glucose, as a result of stress. Stress triggers Gluconeogenesis and Glycogenolysis. *Gluconeogenesis* is a process that converts protein and fat (some fatty acids, but not glycerol) into glucose. Adrenaline from the medulla of the adrenal glands, glucocorticoid (steroid) hormone from the cortex of the adrenal glands, and thyroid hormone from the thyroid gland start gluconeogenesis. *Glycogenolysis* is a process that converts glycogen (the storage of glucose) into glucose. Glucagon, produced by the alpha cells of the pancreas, start glycogenolysis.

Hyperglycemia, from any cause including stress, increases the level of inflammation and other bad results inside the body. The level of stress hyperglycemia is a predictor for the poor outcome of many acute situations, including trauma, stroke, heart attack, and others.

The level of my exercise was strenuous. But, the readings of my blood glucose during the course of running the stairs had not been abnormally hyperglycemic. Perhaps running in the fasting state (early in the morning before breakfast) or my carbohydrate-restricted diet helped prevent hyperglycemia. Of course, I also needed energy for my exercise. My body just did not have cause to produce very much glucose in a short period. If this was true, a person with carbohydrate-restricted diet should have a much lesser chance of hyperglycemia in stressful situations.

Continue To Improve My Health

During this winter, I continued to notice improvement in my general health. Usually in the winter I would catch at least one cold each year. I could get it from my patients who had upper airway infection, when they came to see me. Interestingly after I started the carbohydrate-restricted diet in the fall of 2002, I had had fewer colds. Besides, each one seemed milder than before. I had not needed to use antibiotic for the past five years.

This winter I had only one minor cold, after a patient coughed on me in the middle of examination. I had sneezing and a runny nose with clear, light mucus for three days, then I recovered completely. I did not have to take an antibiotic or any other medications at this time either.

This experience convinced me that the research report about the capability of phagocytosis of the neutrophils, monocytes, and macrophages was valid. *Phagocytosis* is the ability of these white cells to "gobble up" bacteria, viruses, and other foreign bodies. The ability of phagocytosis is decreased, as the level of blood glucose level goes up beyond 100 mg%. Based on the results of my own blood glucose levels, I was very sure that my white cells were stronger than before in defending my body from infections. Besides, keeping my blood glucose level below 150 mg% at all times must be helping to build up my immune system while cutting the level of inflammation. A normal blood glucose level at all times must also improve the collagen fiber of my skin. Not only was the last keratosis on my right thigh continuing to reduce in both hardness and size, but also many old scars were fading away, and becoming as smooth as healthy skin. As somewhat related to the changes of my skin, I had also noticed that I was losing less hair than I had been before I began the carbohydrate-restricted diet. I noticed this simply from the hair quantity found in my shower drain after washing my hairs.

My blood pressure had been very much within the range between 100 and 130 mmHg for the systolic blood pressure, and 50 and 70 mmHg for the diastolic blood pressure. Of course, as I said, blood pressure was dynamic and subject to instant swinging up and down according to the physical and mental state of my body. I was happy as long as my blood pressure did not go up extremely high, when I was excited, and did return to the normal level within a period of 1-4 minutes. I felt this was because I wanted my heart to bear less stress, if possible.

Since June 9, 2002, I continued the frequent monitoring of my heart and carotid arteries with a stethoscope. So far, I still did not hear heart murmurs. I did not hear bruits (the noise as a result of blood flowing through a narrowed carotid artery because of arteriosclerosis or cholesterol plaque) in my carotid arteries either. I had also noticed much less shooting noise in my ears. However, rarely when my blood flow was much stronger, I could hear soft humming sounds in my right carotid artery. That meant that I should continue to monitor my blood volume and blood pressure. In related to the improvement in my cardiovascular system, my ED evaluation continued to improve.

In this season I did not work outside my house as much as I would have liked. My fasting weight did not continue to go down, as I wanted. It had stayed in the range between 157 and 159 pounds with the balance beam scale, which was supposed to be accurate. However, I was not very pleased with the readings. I continued to exercise every day with running the stairs 40 laps, 12 push-ups and 12 sit-ups, and a few repetitions of extending and flexing my back. I could touch the floor with my fingers. This was quite an accomplishment. As in the past, I could not help but questioning whether I would be able to do this kind of exercise five years or ten years from now. I wondered if I would get arthritis in my feet, knees, or hips. Would I be able to maintain current level of stamina or could I make it even better?

This winter, I could not tolerate the cold temperature like I used to, five or six years ago. I realized that after so much lost weight I had very little fat left underneath my skin as an insulator. This was probably the downside of being slimmer. However, I wore only a couple of pieces of thin clothing at home during the winter and I kept the thermostat lower to save energy. I refused to believe that my decreased tolerance of coldness had anything to do with getting older every day. I really hate to blame aging for everything from A through Z.

SPRING 2008

Eve Of The 6th Anniversary Of My Nightmare

June 9th of this year would be the 6th anniversary of my nightmare. In looking back on what I had done to improve my health and save my life, I was very happy and proud of myself. I had found how to change the condition of my body with diet in order to help build up the capability of defense against infections, to prevent from developing diseases, and to promote a slow-down of the aging process.

It is important to have medicines available to us, in the case of a disastrous infection or a rare disease. But it is more important that we keep our body in the best conditions so that we are less likely to depend on medicines very often. We may find that it takes long time to restore the body's system for fighting against infections and diseases. But we may take little or no time to weaken that system by eating more foods rich in carbohydrates, with a high glycemic indices and a high glycemic load.

We should remember that it is never too late to be healthy again. My personal experience of the past six years is an excellent example. Six years ago, no one (including myself) would have believed that I was capable of reforming my health and physical state so much during that period.

Every time I recall my terrible physical condition of six years ago, it still makes me shudder. The memory is like a nightmare from which I cannot wake up. I would still go a long way toward improving and maintaining my health. I will continue my experiments and take notes of my findings, for sharing with people like you.

For example, I frequently wondered if I would be able to continue my rigorous exercise program for five or ten more years from now, or even longer. I also wondered if I would get arthritis in my feet, knees, and hips because of the exercise program. This would be a very interesting experience for all of us including myself. If I succeed in my exploration, we may all be able to follow my path. If I fail, I will let you know about the mistakes I made.

As of this writing, I had sampled more than 1,200 articles and I was still counting. With the findings in these articles, I had learned a lot. I thought that, through my writing, I was able to pass on the information to readers who may be interested in restoring their health by changing their diet.

A Roller Coaster Of My Weight

Since the beginning of March, I had spent more time on writing and editing chapters of this book. I spent very little or no time working outside my home. In addition, I was invited to eat out a few times. I continued to be careful and avoid sugar, flour, grains, and starchy foods. But I could not help but eat some foods prepared by chefs who had used sugar in many of their dishes. As a result, my fasting weight went up to 159 pounds the next morning after each dinner. For the next few days, I had to avoid eating more than my regular diet for dropping my fasting weight. But I had no luck.

My systolic blood pressure during March and the first third of April had been around 140 mmHg or higher when I was relaxed. It was 150 mmHg or higher when I was excited or physically active. I also noticed humming sounds on my right carotid artery when I was excited. That really bothered me very much. I knew I must try harder to lower my fasting weight.

On Friday, April 11, my fasting weight was 158.5 pounds. That day would be sunny and warm, in the upper 70's. I decided to spend as much time as I could to clean my yard. Thanks to my carbohydrate-restricted diet, I did not feel hungry after hard work outside my house all day. I ate my regular meals. The next morning, April 12, my fasting weight went down to 156.5 pounds. Then I worked again for two and a half hours in the morning. At supper, I ate a piece of roasted chicken and a full plate of cold slaw that contained no sugar, but regular ranch salad dressing. The following morning, my fasting weight went up to 157.25 pounds. I was still very worried about my high systolic blood pressure.

On May 8, my wife and I flew to California to visit our son, daughter-in-law, and our beautiful grandson and granddaughter. During our stay, our daughter-in-law and son took us to gardens, a mountain, Lakefront Park, Half-Moon Bay, and the San Francisco seashore. We walked in the gardens and climbed the mountain. They took us to many restaurants for delicious foods. They accommodated me very much by ordering foods with little or no starch, sugar, or flour. On May 17, we attended our granddaughter's piano recital. She played very well. That day was our grandson's birthday. I promised him and our daughter-in-law that I would eat a piece of ice cream cake. And I did.

On May 18, my wife and I flew home. We ate some vegetables and fish a couple of times on our way home. The next morning, my fasting weight was only 158.00 pounds. Not too bad!

After working outside my house, my fasting weight had slowly dropped. On the morning of May 31, my fasting weight was 156.00 pounds. My blood pressure had also come down. I did not hear the shooting noise in my ears. I had also resumed my exercise as soon as we got home. I felt only very mild pressuring in my left upper back occasionally in the exercise. My loaded weight at 11:30 PM was 155.75 pounds. I believed that my fasting weight was getting closer to my goal soon.

My Skin And Keloids

In my journal of six years, I mentioned a few times that my skin conditions had improved a lot. I was still amazed by the positive change in the lesions with keratosis. In my practice, I have seen many patients with keloid formations. Keloid usually appears in an open wound from trauma or surgery. However, there are reports of keloid formations on the skin of people who have had no trauma or surgery. It is an overgrown scar tissue. Sometime, it can grow like a crab. It is bright red in color so it has quite an alarming appearance. We have tried to remove keloid formations with surgery. Most of the time, more surgeries tend to simply give the patient more and bigger new keloids. Based on my old medical knowledge, I believe that keloid must often inflicts itself on African Americans. Asians are second in number. Caucasian people seem to be much less likely to encounter it. Because inflammation probably promotes the build-up of keloid, I have always suspected that people with higher blood glucose level would have higher risk of keloid. Keeping the person's blood glucose level lower by eating low-carbohydrate diet should help prevent and possible reduce keloid formation. At the same time, the same diet probably could reduce tissue adhesion both inside and outside of our body cavities after surgery, such as bowel adhesion with blockage..

Is Cereal Really Good For Us?

Most of us eat cereal for breakfast. The cereal companies claim that cereal could lower our cholesterol, especially LDLs. But I wanted to find out how cereal may affect our blood glucose level. I decided to do an experiment. I asked for three cups of a leading brand of cereal from my elder daughter.

In Nutrition Facts on the box, it suggests one cup (28 grams) for a serving size. In each serving, there are 2 grams of total fat, 0 mg of cholesterol, 190 mg of sodium, 170 mg of potassium, and 3 grams of protein. Also, there are 20 grams of total carbohydrates, including 3 grams of dietary fiber (1 gram of soluble fiber), 1 gram of sugar, and 16 grams of other carbohydrates. So, the net digestible carbohydrates are 18 grams. Total calories for one serving is 100 calories. Not bad, right?

On the morning of May 25, I woke up at 6:54 AM, and weighed 157 pounds. I did not test my fasting blood glucose immediately after getting up. I prepared breakfast as usual. At 9:07 AM, my fasting blood glucose was 101 mg%, before I was to eat my regular breakfast and take multiple vitamins, Vitamin C, and fish oil. At 9:11 AM, I finished my regular breakfast of an egg, 1/6 of medium tomato, 1/3 cup of shred cheese, a small link of fresh sausage, and 1/3 of medium avocado, plus 7 ounces of water. My blood glucose was 101 mg% at 9:21 AM and 100 mg% at 9:37. As expected, my blood glucose after the regular breakfast did not change my blood glucose much. I "swallowed down" a cup of cereal with 7 ounces of water between 9:37 AM and 9:41 AM. My blood glucose was 110 mg% at 9:52 AM, 141 mg% at 10:07 AM, 124 mg% at 10:22 AM, and 106 mg% at 10:37 AM. I went to the restroom at 10:45 AM, and carried a washcloth basket down the stairs at 10:46 AM. My blood glucose was 119 mg% at 10:57 AM. Then, I had a discussion about this experiment with my daughter on the phone for three minutes.

I was somewhat excited. My blood glucose was 135 mg% at 11:07 AM, 148 mg% at 11:22 AM, 149 mg% at 11:37 AM, and 129 mg% at 11:52 AM. During the period, there was no physical activity. At 11:55 AM, I had a BM, then changed my clothes. Again, my blood glucose was 111 mg% at 12:07 PM, 125 mg% at 12:22 PM, and 114 mg% at 12:37 PM. (Please see Figure 31.)

Date: 05/25/2008 Fasting Body Weight: 157 pounds.

Time	6:54	9:07	9:21	9:37	9:41	9:52	10:07	10:22	10:37
BG		101	101	100		110	141	124	106
Note	Awake	FBG	*1		*2	*3	*3	*3	*4
Time	10:57	11:07	11:22	11:37	11:52	12:07	12:22	12:37	
BG	119	135	148	149	129	111	125	114	
Note	*5		*6	*6	*6	*7	*8		

FBG: Fasting blood glucose. *1: Finishing a regular breakfast at 9:11 AM. *2: Eating one cup of cereals with 7 ounces of water between 9:37-9:41 AM. *3: Meeting with guests. *4: Restroom break and carrying a basket of washcloths downstairs. *5: Discussing with my elder daughter at 10:53-10:56 AM. *6: No physical activity. *7: BM and changing cloths. *8: Working with a computer.

Figure 31. Cereals and My Blood Pressure and
Blood Glucose Levels (1) on 05/25/2008

From the result above, I was surprised at that how fast and how much my blood glucose increased, after eating one cup of cereal. Luckily, my regular carbohydrate-restricted breakfast probably offset the glycemic index of the cereal. That was why my blood glucose was not up so much in its first peak. After my small intestine took up the regular carbohydrate-restricted breakfast, the rest of the cereal made the second peak in my blood glucose. But, I had to give a benefit of doubt in this case. The second peak in my blood glucose could have been caused by stress hyperglycemia from the exciting discussion with my elder daughter. She is a medical doctor. The only problem was that the "exciting effects" probably lasted too long. After all, I was impressed with the effects on my blood glucose by only a cup of cereal. I expected the impact would be more extreme if I had added fruit and milk to a bowl of cereal. My daughter suggested that I should have eaten the cereal before my regular breakfast. I would do just that tomorrow. The next morning, May 26, I woke up at 7:00 AM, and let my dog go out. My fasting weight was 156.25 pounds. My fasting blood glucose was 100 mg% at 7:12 AM. At 7:21 AM, my blood pressure from my right arm was 107/56 mmHg with heart rate at 71 per minute. At 7:41 AM, my second fasting blood glucose was 101 mg%. I immediately ate a cup of cereal with 7 ounces of water. I finished it at 7:45 AM. At 7:56 AM, my blood glucose was 116 mg%. In the meantime, I was preparing my family regular breakfast. That should help spend some energy from the cereal. My blood glucose was 161 mg% at 8:11 AM, 180 mg% at 8:26 AM, 166 mg% at 8:41 AM, 113 mg% at 8:56 AM, 112 mg% at 9:11 AM, and 106 mg% at 9:26 AM. At this point, my pancreas took good care of the glucose from the cereal.

I started my regular breakfast, plus multiple vitamins, Vitamin C, and fish oil, at 9:26 AM; and finished them at 9:30 AM. My blood glucose was 102 mg% at 9:49 AM and 105 mg% at 10:05 AM. Again, the readings showed that my regular breakfast did not raise my blood glucose. I stopped my experiment, because I had to go out to my office at 10:20 AM. When I returned home at 12:08 PM, my blood glucose was 115 mg%. (Please Figure 32.)

Date: 05/26/2008 Fasting Body Weight: 156.25 pounds.

Time	7:00	7:12	7:21	7:41	7:56	8:11	8:26	8:41
BG		100		101	116	161	180	166
Note	Awake	FBG	*1	*2	*3			
Time	8:56	9:11	9:26	9;49	10:05	10:20	12:08	
BG	113	112	106	102	105		115	
Note						*4	*5	

FBG: Fasting blood glucose. *1: Blood pressure from right arm at 107/56 mmHg, heart rate at 71/minute. *2: Eating one cup of cereals with 7 ounces of water between 7:41-7:45 AM. *3: Preparing breakfast. *4: Going to office. *5: Returning home.

Figure 32. Cereals and My Blood Pressure and
Blood Glucose Levels (2) on 05/26/2008

From the results, I was even more surprised at how much the effect on my blood glucose by only a cup of cereal. My blood glucose went up to 180 mg% in 40 minutes! I was not sure if the readings could be higher than 180 mg% in between 8:11 AM and 8:41 AM. Luckily, my pancreas handled the increase of blood glucose well. My blood glucose returned to normal range about one hour later. If I had added banana, apple or other fruits, plus milk, especially the skim milk, my blood glucose would have gone even higher for a longer period. To the best of my knowledge, this was unacceptable to me! By the way, the label "whole grains" does not impress me, not even to question whether "whole grains" really cut down our cholesterol. The food manufacturers should produce foods with a much lower in glycemic index and glycemic load, and higher fibers. They should not add sugar to them!

Update Of My Diet

October 12, this year, it will be the 6th anniversary of changing my diet to carbohydrate-restricted diet. Often,

I looked back at what I had accomplished with my diet for more than five and a half years. Always, I was very grateful to my excellent decision to stop playing with water restriction and no-salt diet.

After switching my regular high carbohydrate diet to carbohydrate-restricted diet, I had to eat more meat and pork, as well as vegetables at first. That was because I had to fill the volume, which used to fill with rice, flour foods, and starch. As the time went on, I had no feeling of hunger or sickness from hypoglycemia. Without having to remind myself, I had eaten less in volume and calories. I was always satisfied with the foods on the table. And, I ate because of the mealtime, not because of hunger.

The basics in designing my diet was (1) To keep the amount of carbohydrates as small as possible, but avoid forming ketone in my body; (2) To eat three meals daily, if possible; (3) to eat more calories for breakfast, more fibers for lunch, and less calories for supper; (4) To have supper early, if possible, and no snack after supper. (Please see Figure 33.)

Nutrition Facts of My Breakfast and Lunch

Breakfast:	Serv. No.	T. Cal.	Fat Cal.	Fat	Sat. Fat	Choles.	Sodium	Carbo.	Fiber	Sugar	Protein
Eggs	3.00	240.00	135.00	15.00	4.50	0.72	0.21	4.00			21.00
Peanut Oil	6.00	780.00	780.00	84.00	15.00	0.00	0.00	0.00			0.00
Stella Cheese	1.30	117.00	78.00	9.10	5.85	0.03	0.25	1.30	1.30	0.00	0.00
Sausage	1.00	160.00	110.00	12.00	4.00	0.04	0.70	1.00	0.00	0.00	10.00
Tomato	0.50	11.16						1.69			
Avocado	1.00	306.00	270.00	30.00				12.00	8.50	3.50	3.70
Soybean Oil	6.00	780.00	780.00	84.00	12.00	0.00	0.00	0.00			0.00
Total with P		1,614.16	1,373.00	150.10	29.35	0.78	1.16	19.99	9.80	3.50	34.70
Total with S		1,614.16	1,373.00	150.10	26.35	0.78	1.16	19.99	9.80	3.50	34.70
My Share P		664.19	567.20	62.01	11.98	0.33	0.43	8.31	4.12	1.47	13.71
My Share S		664.19	567.20	62.01	10.72	0.33	0.43	8.31	4.12	1.47	13.71
Lunch											
Tuna	2.50	175.00	25.00	2.50	0.00	0.06	0.63	0.00			37.50
Salmon	2.50	200.00	67.50	7.50	3.75	0.03	0.43	0.00			30.00
Chicken	6.00	360.00	30.00	3.00	0.00	0.18	1.20	0.00			78.00
C Cheese	1.50	150.00	135.00	15.00	9.00	0.05	0.14	1.35		1.35	3.00
Mayonnaise	8.00	800.00	720.00	80.00	12.00	0.04	0.72	0.00			0.00
Tuna Mix		1,125.00	880.00	97.50	21.00	0.15	1.48	1.35	0.00	1.35	40.50
Salmon Mix		1,150.00	922.50	102.50	24.75	0.11	1.28	1.35	0.00	1.35	33.00
Chicken Mix		1,310.00	885.00	98.00	21.00	0.27	2.06	1.35	0.00	1.35	81.00
Celeries (6")	4.00	1.32						0.72			
Romain Lett	4.00	6.80						1.32			
My Share TC		451.32	352.00	39.00	8.40	0.06	0.59	1.26	0.00	0.54	16.20
My Share TR		456.80	352.00	39.00	8.40	0.06	0.59	1.86	0.00	0.54	16.20
My Share SC		461.32	369.00	41.00	9.90	0.04	0.51	1.26	0.00	0.54	13.20
My Share SR		466.80	369.00	41.00	9.90	0.04	0.51	1.86	0.00	0.54	13.20
My Share CC		437.99	295.00	32.67	7.00	0.09	0.69	1.17	0.00	0.45	27.00
My Share CR		443.47	295.00	32.67	7.00	0.09	0.69	1.77	0.00	0.45	27.00

Legends: **T. Cal.**: Total Calories. **Fat. Cal.**: Calories from fat. **Fat**; All fats. **Sat. Fat**: Saturated fat. **Choles.**: Cholesterol. **Carbo.**: Total carbohydrate. *All individual nutrients weighed in grams.* **Total with P.**: The total figures when the breakfast uses peanut oil. **My share P.**: The total figures for my breakfast, which uses peanut oil. **Total with S.**: The total figures when the breakfast uses soybean oil. **My share P.**: The total figures for my breakfast, which uses soybean oil. **Tuna Mix**: The lunch mix uses Tuna. **Salmon Mix**: The lunch mix uses Salmon. **Chicken Mix**: The lunch mix uses chicken. **My Share TC**: Total figures for my lunch with Tuna Mix and celeries. **My Share TR**: Total figures for my lunch with Tuna Mix and Romaine lettuce. **My Share SC**: Total figures for my lunch with Salmon Mix and celeries. **My Share SR**: Total figures for my lunch with Salmon Mix and Romaine lettuce. **My Share CC**: Total figures for my lunch with Chicken Mix and celeries. **My Share CR**: Total figures for my lunch with Chicken Mix and Romaine lettuce.

Figure 33. Update of My Diet (In this chart, I did not include the amount of fibers from celery and lettuce.)

Based on our menu, I took in about 650 calories from breakfast, and about 450 calories from lunch. Because of the variety in the menu for supper, I could not exactly calculate the amount of calories for supper. However, I must have taken in between 400 calories and 600 calories for a regular supper. My total daily calories were between 1,500 calories and 1,900 calories, depending on the snacks, including nuts and fruits. In these I had reduced the amount of daily calories from carbohydrates; I left the amount of fats and proteins about the same as I did before starting my diet. The most important experience with the diet was that I had never felt hungry, even though I had taken in fewer calories than before.

Chapter 6: Carbohydrates Can Kill

A MOUNTAIN OF EVIDENCE

I told you about my personal experience in Chapter 5. Now, I want to show you a large volume of research projects and reports; they support my experience and findings. (Please visit www.carbohydratescankill.com for the reading list. The articles with my notes in this chapter are the first articles in each category on the list.) The most shocking findings in these research reports draw link between the common practice of overeating carbohydrates and the causes of most diseases, which people suffer during their lives.

Eating too many carbohydrates not only makes us gain more weight and become obese. It also sets our body up for different diseases. These diseases range from inflammation and infection to heart diseases, stroke, Alzheimer's disease, and cancers. This is why I am so eager to share with you my personal experience and to give you the information about a small fraction of the research articles that I have reviewed. I want you to know that eating fewer carbohydrates is very important to maintaining good health.

EXCESS WEIGHT AND OBESITY

Overweight/Obesity And Osteoarthritis

Marks R and Allegrante JP. "Body mass indices in patients with disabling hip osteoarthritis." <u>Arthritis Research & Therapy.</u> Volume 4, Number 2, Pages 112–116. 2002. http://www.pubmedcentral.nih.gov/articlerender. fcgi?artearch.id=83842

One of the harmful impacts of excess weight/obesity is to destroy the structure of our joints with inflammation. This most often happens in our hips and knees, which bear our weight whenever we are on our feet. One of the common problems for our joints is osteoarthritis. Osteoarthritis is a result of inflammation of the bony part of a joint. With inflammation, the joint is eventually damaged out of its original shape, and becomes painful.

We use Body Mass Index (BMI) to determine if we are overweight or obese. Please see Figure 19, Body Mass Index, in Chapter 2 for your quick reference. One is underweight if his BMI is 18 or less; he is within normal weight if his BMI is between 18.1 and 24.9; he is overweight if his BMI is between 25 and 29.9; and he is obese if his BMI is 30 and over.

This study found the link between hip osteoarthritis and BMI in a group of men and women, from 23 to 94 years of age who had already required replacement of their hip joints. The study found that overweight men and women have a high risk of having end-stage arthritis of the hipbone. Obese patients with hip joint replacement have a higher chance of requiring replacement of the initial hip prosthesis sometime in the future.

Weight Gain And Sleep Apnea (Airway Obstruction)

Rubinstein I, et al. "Paradoxical glottic narrowing in patients with severe obstructive sleep apnea." Journal of Clinical Investigation. Volume 8, Number 4, Pages 1051–1055. April 8, 1988. http://www.pubmedcentral.nih.gov/articlerender.fcgi?artid=329630

Excess weight and obesity can boost the odds of collapsing our airway or sleep apnea. *Apnea* means that there is no air flowing in and out of the airway. The anatomy of sleep apnea or airway obstruction has already been described in Chapter 2.

This study concluded that the irregular collapses of the glottis (the entrance of the larynx) were the cause of OSA (obstructive sleep apnea) for the obese men. (Please see AIRWAY OBSTRUCTION, chapter 3.) This was true even if the ability of their pharyngeal space to collapse was at the low end of the normal range.

Obesity, Sleep Apnea, And High Blood Pressure

Wolk R, et al. "Obesity, sleep apnea, and hypertension." Hypertension, Volume 42, Number 6, Pages 1067-74. December 2003. http://hyper.ahajournals.org/cgi/content/full/42/6/1067

What is the consequence, if one suffers repeated airway obstruction? Breathing brings oxygen in the air to the blood, which is circulating through the lungs; it is also to bring out carbon dioxide from the blood to outside of the body. With airway obstruction, the blood cannot get oxygen that the body needs; and it cannot empty carbon dioxide built up in the blood and the lungs. Without the supply of oxygen, our cells or tissues would die quickly, especially our brain and nerves, heart, and other vital organs. Also, a build-up of carbon dioxide is harmful to our organs. A pool of physiological mechanisms interacts among obesity, obstructive sleep apnea, and hypertension. This study suggested that coexistence of obesity and obstructive sleep apnea might destroy the cardiovascular system and it might also contribute to some of the "metabolic syndrome." Metabolic syndrome is a combination of cardiovascular diseases and complications of diabetes mellitus. Clinically, patients with both obesity and obstructive sleep apnea complain of "poor sleep quality, widespread apnea, considerable daytime sleepiness", and so on. These symptoms alone have already seriously affected the quality of the patients' life.

Obesity, Sleep Apnea, And Atrial Fibrillation

Gami, AS, et al. "Obstructive Sleep Apnea, Obesity, and the Risk of Incident Atrial Fibrillation." Journal of American College of Cardiology. Volume 49, Number 5, Pages 565-571. 2007. http://content.onlinejacc.org/cgi/content/abstract/49/5/565?ck=nck

Our heart has four chambers, two for the top, and two for the bottom. The top two chambers are atria and the bottom two chambers are ventricles. After circulating the whole body, blood flows back through the largest vein, vena cava, into the right top chamber, or right atrium. The heart muscles contract in waves from the atrium to the ventricle. At each wave of contraction (or each heartbeat), blood is squeezed from the right atrium into the right bottom chamber or right ventricle. Then the blood flows out of the right ventricle into the pulmonary artery to circulate through the lungs. In lung circulation, the blood releases carbon dioxide into and takes in oxygen from the alveoli (the space at the end of respiratory tract). The fresh blood, loaded with oxygen from the lungs, enters the left atrium through the pulmonary vein. In the same heartbeat mentioned above, the blood is squeezed from the left atrium into the left ventricle. Then the blood flows out of the left ventricle into the aorta, the largest artery of the body, for circulating through the whole body. The blood is moved through the arteries in waves generated by the contracting heart, which is supposed to beat regularly. What happens if the heart does not beat regularly?

There are several types of irregular heart rhythms, depending on their causes. Atrial fibrillation is one of the irregular rhythms; its diagnosis is made when the muscles of the atria do not contract forcefully and regularly, but weakly, irregularly, and rapidly instead. So, the atria cannot effectively push the blood into the ventricles that are also beating irregularly and faster than a normal pace.

In the case of atrial fibrillation, blood in the atria tends to flow around inside the chamber and to develop blood clots. When the blood clot is pushed out of the atria into the ventricles, it can cause pulmonary embolism if it enters the lungs. It can cause stroke if it enters the brain. It can cause heart attack if it enters the coronary artery. It can also cause blockage of arteries to other vital organs or limbs if it reaches the organs or limbs. Preventing or treating atrial fibrillation is important. If untreated, its consequences are serious.

This study found that obstructive sleep apnea is an independent risk factor for developing atrial fibrillation among patients who are under 65 years of age. The odds were 2.18 times higher in those with obstructive sleep apnea than those without. Also, those with low blood oxygen concentration during sleep would have a 3.29 times higher chance of developing atrial fibrillation than those with a normal blood oxygen concentration.

Heart failure was the major risk factor among patients 65 years and older who had developed atrial fibrillation. The researchers also found a strong correlation between the risk of atrial fibrillation and obesity alone. It is an excellent suggestion that we can reduce the risk of obstructive sleep apnea by keeping our weight down!

Obesity Shortens Life

Fontaine KR, et al. "Years of Life Lost Due to Obesity." Journal of American Medical Association, Volume 289, Pages 187-193. 2003. http://jama.ama-assn.org/cgi/reprint/289/2/187.pdf

Public health officials and organizations have alerted Americans about the danger of obesity. Nonetheless, the population has continued to grow heavier. Obesity is often linked to various harmful diseases that shorten the lives of its victims. This study concluded that the degree of one's obesity (BMI) has a close link to the number of years of his life are to be lost. It also stressed that obesity in one's younger age will result in many more years of his life expectancy lost.

As of this writing, we have witnessed the rapid spread of weight gain/obesity in the young children. Parents, pediatricians, and public health officials should take the necessary steps in changing children's diets, for improving their health and life expectancy.

Obesity Is A Global Problem

Bertsias G, et al. "Overweight and obesity in relation to cardiovascular disease risk factors among medical students in Crete, Greece." BMC Public Health. Volume 3, Number 3. 2003. http://www.pubmedcentral.nih.gov/picrender. fcgi?artid=140012&blobtype=pdf

Obesity has also had a global outbreak. It grows in the developed and developing countries alike, wherever food supply is plentiful.

For your reference, a body mass index between 25.0 and 29.9 is overweight; and 30.0 or over is obesity. Central obesity is the buildup of fat in the abdominal portion (belly); it is also a risk factor for cardiovascular disease.

A study of Greek medical students found that about 40% of the male students and 23 % of the female students were overweight or obese. There were 33.4% of the male students and 21% of the female students suffered central obesity (large belly). There was a strong link between hypertension and a high BMI. Abnormally high lipoprotein values were found in both the male students who had a larger ratio on the waist circumference versus their height and the female students who had a larger waist circumference.

CARBOHYDRATES AND OBESITY

We Already Knew About High Fats Diet For Weight Loss

Farquhar JW, et al. "Glucose, Insulin, Triglycerides Responses to High and Low Carbohydrate Diets in Men."

Journal of Clinical Investigation, Volume 45, Number 10, Pages 1648-1656. 1966. http://www.pubmedcentral.nih.gov/picrender.fcgi?artid=292847&blobtype=pdf
Sometimes I have wondered if health professionals have been paying attention to the research reports. Many research results contradict the ideas they have promoted for a long time. In 1966, a study found an association between the high carbohydrate diet and the increase of serum triglyceride, cholesterol, and phospholipids (molecules including both lipid and phosphate). The study also accidentally observed weight loss with low-carbohydrate, high-fat diets. Based on the results of this study, researchers had already questioned the logic behind the general recommendation of high-carbohydrate-low-fat diets for the prevention of coronary heart disease. After reviewing this report, we should ask an interesting question. Why are our medical and nutritional professionals still unwilling to redesign a sensible dietary recommendation for us?

Calorie Count Is Not The Most Important Factor In Weight Loss

Feinman RD. "'A calorie is a calorie' violates the second law of thermodynamics." Nutritional Journal. Volume 3, Number 9. July 28, 2004. http://www.pubmedcentral.nih.gov/articlerender.fcgi?tool=pmcentrez&artid=506782
The disparity between the estimated and the actual amounts of calories in a diet has been a myth. Until now, there are still only a few researchers who understood and recognized that.

Our body can use carbohydrates to produce energy by breaking them down all the way to water and carbon dioxide.

The process wastes little energy in producing heat. However, our body cannot use protein to produce energy without having to convert protein into carbohydrate first. Converting protein and fat for producing energy (gluconeogenesis) is triggered when the body is short of glucose supply. (Please see DIGESTION AND METABOLISM in Chapter 1)

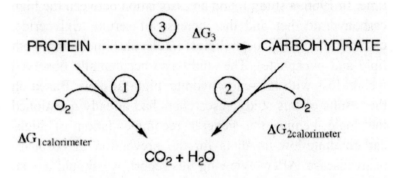

Pathways for oxidation of macronutrients.
Source: Figure 1. "A calorie is a calorie" violates the second law of thermodynamics."
Reprint with permission from Professor Richard Feinman

The study uses this diagram (the original figure) to indicate that in both Process 1 and Process 2 each gram of proteins and carbohydrates would supposedly produce 4 kilocalories of energy. At the same time, protein and carbohydrate are converted into carbon dioxide and water. However, protein cannot be converted and produce energy through the process 1 in our body. Instead, protein has to become carbohydrates first for producing energy, through process 3; there is an expense of energy, ΔG_3, for the conversion. At the end of process 3, the converted carbohydrate from protein will hold about 3 kilocalories.

Each gram of protein will then provide our body about 3 kilocalories of energy when it goes through the process 2. The lost energy becomes heat (thermic effect). The thermic effect is 2-3% for fat, 6-8% for carbohydrates, and 25-30% for protein. From this article, it should be very clear that the kind of foods (carbohydrates, fats, and proteins) which we eat matters more in controlling our body weight than the amount of calories we consume.

Greater Weight Loss With The Low-Carbohydrate Diet

Foster GD, et al. "A Randomized Trial of a Low Carbohydrate Diet for Obesity." New England Journal of Medicine, Volume 348, Number 21, Pages 2082-2090. May 22, 2003. http://content.nejm.org/cgi/content/full/348/21/208 2?ijkey=03ddafab832a81a30330c9ae0ed4112e986169bd

It is true that consuming fewer calories than the body needs for performing its daily activities will result in weight loss. However, it is not easy for many of us to cut back on daily calories because it is somewhat similar to starving ourselves.

There are numerous dietary formulas, chemicals, and foods being suggested for weight loss. Many of them do not work. Others stir up debates. One of the most debated plans uses a low- carbohydrate diet for weight loss. Most of us misunderstand thermodynamic laws. At the same time, they do not have the information about the long-term impact on our health, with a low-carbohydrate diet. The debate will continue. This study found both the low-carbohydrate, high-protein, high-fat and the *low-calorie*, high-carbohydrate conventional diets promote weight loss. LDL was significantly lower with the low-calorie conventional diet group at the end of the first three months. However, the low-carbohydrate diet increased the concentration of high-density cholesterol (HDL) and it lowered the concentration of triglycerides. Both diets lowered the diastolic blood pressure and the insulin response to the glucose challenge by mouth (taking in 75 grams of glucose solution by mouth). There was improvement in sensitivity to insulin with both diets during the study.

The study also found the benefits of the low-carbohydrate diet in greatly improving some of the risk factors for coronary heart diseases. Also, they found that many participants had difficulties in adhering to the diets and dropped out the study. Nevertheless they suggested that longer and larger studies were needed to clarify the long-term safety and efficiency of the low-carbohydrate, high-protein, and high-fat diet. From this report, we can appreciate that dropping the amount of carbohydrates in our diet, even just a small drop, helps prevent diabetes (DM) or make it more manageable. The higher dropout rate with a low-calorie conventional (high-carbohydrate) diet group meant that weight loss by starvation is a workable but difficult choice.

Very-Low-Carbohydrate Diet vs. Calorie-Restricted Low-Fat Diet

Brehm BJ, et al. "A Randomized Trial Comparing a Very Low Carbohydrate Diet and a Calorie-Restricted Low Fat Diet on Body Weight and Cardiovascular Risk Factors in Healthy Women." Journal of Clinical Endocrinology and Metabolism, Volume 88, Number 4, Pages 1617-1623. April

2003.　　http://jcem.endojournals.org/cgi/content/full/88/4/
1617?ijkey=5d05289221d184497af60b95e961da6abb5f
b647

Debates continue on the controversy about the efficacy and safety of using low carbohydrate diets in stopping and reversing the national trend of obesity. Numerous studies have been or are being conducted in search of answer(s). This study used a very low carbohydrate diet with 20 grams of carbohydrates, and up to 40-60 grams of carbohydrates for participants whose urine contained ketone. It also used a *calorie-restricted*, moderately low-fat diet in which 55% of its calories came from carbohydrates, 15% from proteins, and 30% from fats. Ketone is a product of metabolizing carbohydrate, fat, and protein by our body. (Please see Figure 5. Metabolism of the Nutrients.)

The study found more weight loss (nearly 2:1) with the very low carbohydrate dieters than with the calorie-restricted low fat dieters. The product, ß (beta)-hydroxybutyrate, found when using fats as the source of energy, was remarkably increased in the plasma (the fluid part of the blood), in the very low carbohydrate dieters at the end of the first three months in the study. The loss of fat mass and lean body mass were found more pronouncedly in the very low carbohydrate dieters than the calorie-restricted low fat dieters. There was *no mineral depletion* found in either group. Both groups had significant decreases in total cholesterol, low-density lipoprotein cholesterol (LDL-c), and triglycerides at the first three-month assessment. There was a significant increase in high-density cholesterol at the first six months; there were no changes in blood pressure in either group.

The study concluded that a very low carbohydrate diet

is more effective than the calorie-restricted low fat diet for short-term weight loss over six months' time. It was not associated with any harmful effects, such as critical cardiovascular factors in healthy women. As in some of other studies, when placing restriction on the total calories of the conventional diets, a reduced amount of carbohydrates in the diet is always expected to improve the picture of the lipids (fats) of the blood in the short-term. Let us see some studies that compare diets with the same amount of calories.

The Very-Low-Carbohydrate Diet vs. The Low-Fat Diet

Volek JS, et al. "Comparison of energy-restricted very low-carbohydrate and low-fat diets on weight loss and body composition in overweight men and women." Nutrition and Metabolism, (London), Volume 1, Number 13. November 8, 2004. http://www.pubmedcentral.gov/articlerender.fcgi?tool =pmcentrez&artid=538279

More recent studies have shown a strong correlation between the fat mass of our waistlines and the risks for cardiovascular disease, cancers, diabetes (DM), and other serious diseases. We want to learn the effects on weight loss with different diets. We also want to know the impact of different diets on our body composition as well as how the diets may affect fat mass in the waistline. This study showed definite advantages in a low-caloric, very-low-carbohydrate diet over the low-caloric, low-fat diet for men in weight loss (both total and waistline), even with the higher consumption of energy by those dieters. Most women also responded positively to the very-low-carbohydrate diet, particularly in terms of waistline fat loss. In both men and women, the significant loss of waistline fat is not only part of the total fat loss, but the ratio of the waistline fat to the total fat is also especially decreased. In another words, the waistline was changed from the barrel shape into the pear shape.

The study affirmed the clear advantage of a very-low-carbohydrate diet over a low-fat diet in promoting weight loss, both in total fat loss and in waistline fat loss. Most researchers agree that it is important that more studies be conducted for confirmation of the degree of waistline fat loss in women using this diet.

A Low-Carbohydrate Diet vs. A Low-Fat Diet In Severe Obesity

Samaha FF, et al. "A Low-Carbohydrate as Compared with a Low-Fat Diet in Severe Obesity." New England Journal of Medicine, Number 21, Volume 348, Pages 2074-2081. May 22, 2003. http://content.nejm.org/cgi/content/abstract/348/21/2074

More recent studies revealed a positive connection between obesity and many health problems. We are urgently looking for ways to help obese people lose weight. There are promotions everyday on new exercise programs, dietary formulas, lifestyle modifications, and others. However, many of them do not work as claimed.

This study observed that the average weight loss was 5.8 kilograms for the low-carbohydrate dieters and 4.2 kilograms for the *calorie-restricted* low-fat dieters. The averaged decrease in triglycerides was 20% for the low carbohydrate dieters and only 4% for the calorie-restricted low fat dieters. The decrease did not correlate to whether the participant was previously treated for diabetes and/or high cholesterol.

The study found improvement in insulin sensitivity at an average of 6% for the low carbohydrate dieters; and 3% for the calorie-restricted low fat dieters. This study concluded that severely obese patients, who had diabetes mellitus or metabolic syndrome, would lose more weight with a low carbohydrate diet than with a calorie-restricted, low-fat diet. The same was true of the comparative improvement in insulin sensitivity and triglyceride levels.

Metabolic syndrome, previously called syndrome X, is a disorder in which the body does not respond to insulin secreted by its pancreas despite a high level of insulin in its circulation. Also, there is a high level of blood glucose (sugar) associated with insulin resistance or an advanced degree of diabetes (DM). A high level of blood sugar causes different disease processes including heart and blood vessel problems, brain and nerve disorders, and cancers.

More Loss In Water With Low-Carb Diet Than Low-Fat Diet

Yang M and Van Itallie TB. "Composition of weight lost during short-term weight reduction. Metabolic responses of obese subjects to starvation and low-calorie ketogenic and nonketogenic diets." Journal of Clinical Investigations, Volume 58, Number 3, Pages722–730. September 1976. http://www.pubmedcentral.gov/articlerender.fcgi?tool=pmce ntrez&artid=333231

More studies have shown a better result in losing weight with low-carbohydrate diets. However, many researchers have criticized these diets for causing more loss of water than fat. They also expressed concern about the production of ketone bodies by the very low carbohydrate diet. (Please see KETOGENIC DIET, Glossary of Important Terms, Chapter 1.)

One study investigated the changes in the body composition after either (1) essentially withstanding starvation, (2)) on an 800-kcal ketogenic (low-calorie, low-carbohydrate and high-fat) diet or (3) on an 800-kcal non-ketogenic (low-calorie, mixed) diet. All the diets for this study were calorie-restricted for weight loss. The average weight loss was the greatest from starvation, moderate from the ketogenic diet, and the least from the non-ketogenic diet.

Earlier studies have implied that high fat diets induce both sodium and water loss. This observation should make us rethink the relationships among blood glucose, sodium, and water retention, in cases of hypertension. This is why weight loss can help bring down the blood pressure. Is it logical to use diuretics to drive the water and sodium out, without first looking into the level of blood glucose?

Dietary Fat Does Not Equal Body Fat

Willett WC and Leibel RL. "Dietary fat is not a major determinant of body fat." American Journal of Medicine, Volume 113, Supplement 9B, Pages 47S-59S. December 30, 2002. http://www.ncbi.nlm.nih.gov/sites/entrez?db=PubMed &cmd=Retrieve&list_uids=12566139

It is unfortunate to point out that most people do not have a good understanding of how foods are used by the body. A majority of us still think that fats in our diet make us fat. As a result of the misunderstanding, we have suffered "lipophobia." (I made up this word meaning "afraid of fats.") This makes many of us keep fats as far away from our diet as possible.

- Consider how the population looks today. Thanks to the Advisors from public health agencies, including both medical and nutritional professionals, have conducted a tireless campaign to promote eating more carbohydrate and little fat for more than half a century. The population has grown heavier every year, not only in the United States, and also in most other part of the world.

This review article focused on biochemical and physiological viewpoints, studies on metabolism for an intermediate period, long-term observation reports, trials for weight loss and experiments for attempting to reduce the amount of body fat. The article also examined the differences in dietary fats and body fat among different countries and populations, as well as in various US population groups. They reviewed the relationship between genetic factors and body fat while pursuing other theories about how differing bodies regulate fat mass.

The fact that the migrants to the United States a century ago gained so much weight in a short period of time puts the genetic factor into question. Furthermore, the consumption of fats in the United States had been on the decline for two recent decades, as of the time this review was published. During the same period, as the population grew heavier, the link between the dietary fats and body fat was further discredited. This review concluded that there were still questions about the mechanism that creates body fat. At least they suggested that dietary fats were not the major determinant of body fat.

Fat Is Not Fattening, If Carbohydrates Are Restricted

Feinman RD. "When is a high fat diet not a high fat diet?" Nutrition & Metabolism (London). Online October 17, 2005. http://www.nutritionandmetabolism.com/content/2/1/27

The dietary modification for weight loss includes calorie-restricted high-carbohydrate low-fat diets, low-carbohydrate high-fat diets, and low-carbohydrate high-protein diets, and so on. "High" and "low" are a description of one thing in comparison to another or others. They are not absolute in describing quantities. Using the names of the diets above often causes misunderstanding of the study results, and can likely be misleading.

In an editorial comment on an article in the journal Nutrition and Metabolism, in 2005, author Professor Feinman wrote, "A ketogenic diet reduces amyloidal beta 40 and 42 in a mouse model of Alzheimer's disease". The ketogenic diet in this particular research was a high-fat, very low-carbohydrate diet, with less than 1 % of calories from carbohydrates and 80% of calories from fats. The diet was intended to produce ketones inside the body when it uses fat for the source of energy, under the circumstance of carbohydrate shortage. This article can be found in the Nutrition and Metabolism (London) 2005, 2:28 or online at http://www.pubmedcentral.nih.gov/articlerender.fcgi?tool=p ubmed&pubmedid=16229744

Professor Feinman pointed out that fat was like a bomb, yet played a relatively passive role, while carbohydrate was like a fuse, and played an active role in modulating the role of fat. He further suggested replacement of the principle that "you are what you eat," with one that " you are what you do with what you eat." In the current teaching, we blame excessive free fatty acid for causing insulin resistance. However, the opposite could be logical, stating, as follows:

- that insulin resistance was a result of hyperinsulinemia (too much insulin in the blood circulation);
- that excessive amount of carbohydrate intake triggered too much insulin secretion;
- that the inability to use blood glucose caused increased lipolysis with excessive free fatty acid in the circulation. Lipolysis is breaking the body fat down into fatty acids and glycerol.

Feinman used the similarity in the rate of lipolysis in cases of starvation and carbohydrate-restricted diet. In both situations, there was no carbohydrate to stimulate insulin secretion; in turn, there was no insulin to suppress lipolysis.

He took up the case of dietary manipulation for weight loss. He pointed out that it was the amount of carbohydrates, not its percentage of the total dietary calories that mattered the most in determining whether the body would save or spend fat. He cited a research article by J. S.Volek, who stresses that replacing unsaturated fats with carbohydrates in diet was found to increase the risk of cardiovascular diseases.

Feinman also used a study by M.C. Gannon ("Effect of a high-protein, low-carbohydrate diet on blood glucose control in people with type 2 diabetes," Diabetes. Volume 53, Number 9, Pages 2375-82 September 2004) to underscore his belief that fat takes a passive role in the presence of carbohydrate. In Gannon's research, the level of the patient's blood glucose was decreased to a stable, normal level, with the high-protein, low-carbohydrate (high-fat) diet. The test diet ratio for carbohydrate/protein/fat was 20%/30%/50%. In comparison, that of the control diet was 55%/15%/30%, respectively.

GLYCEMIC INDEX AND GLYCEMIC LOAD

Defining Glycemic Index And Glycemic Load

Foster-Powell K, et al. "International table of glycemic index and glycemic load values: 2002." American Journal of Clinical Nutrition, Volume 76, Number 1, Pages 5-56, 2002. http://www.ajcn.org/cgi/reprint/76/1/5

We often hear people argue, "You cannot compare apples to oranges." Indeed, they are different kind of fruits. However, both of them contain carbohydrates and we can still compare them based on the quality and quantity of their carbohydrate content.

Consistent charts of glycemic index (GI) in scientific publications have been critical in describing the relationship between GI, Glycemic load (GL) and health. (Please see Glycemia, Glossary of Important Terms, Chapter 1) Using GI makes it easier to recognize each food's potential for raising the postprandial (after meal) blood glucose without having to identify its chemical structure (simple, complex, sugar, starch, etc.) as available or unavailable carbohydrate. The same types of foods can yield various values of GI, depending on the differences in their ingredients, the methods of preparation, and so on.

After eating a number of units of a carbohydrate food, we can estimate the effect of the food on the body by multiplying its glycemic index and the number of units of that food eaten. Their product is glycemic load (GL). Glycemic load represents the quantity of the food or foods. Long-term use of foods with high GL makes one more susceptible to type 2 diabetes and coronary heart disease.

Low GI Diet In Treatment Of Pediatric Obesity

Spieth LE, et al. "A low-glycemic index diet in the treatment of pediatric obesity." Archives of Pediatric Adolescent Medicine. Volume154, Number 9, Pages 947-51. September 2000. http://archpedi.ama-assn.org/cgi/content/full/154/9/947

The overweight population is growing younger. Obesity in childhood is a major pediatric problem that causes diseases in both childhood and later in the adulthood. It can shorten the obese child's life expectancy. We critically need solutions for stopping this trend. Public health agencies and medical professional organizations emphasize that children must spend more time outdoors exercising, and less time indoors watching television or playing video games. They also call for changes in children's nutrition. However, there is a serious question as to whether the recommended amount of daily carbohydrates consumed is really healthy.

This study compared the effectiveness in weight loss between a balanced, energy-restricted, reduced-fat diet and a low-GI (glycemic index) diet. The balanced, energy-restricted, reduced-fat diet limited the use of high-fat, high-sugar, and energy-dense foods. It increased the use of grain products, vegetables, and fruit. This diet consisted of carbohydrates (55-60%), proteins (15-20%), and fats (25-30%). The "Low-GI Pyramid" model placed vegetables, fruits, and legumes at its bottom, lean protein and dairy products on the second tier, whole-grain products on the third, refined grain products on the fourth, and sugars at the top. The low-GI diet itself consisted of carbohydrates (45-50%), proteins (20-25%), and fats (30-35%). The study observed that the average of body mass index (BMI) decreased, 1.53 for the low-GI group and 0,06 for the low-fat group. The average body weight was a decrease of 2.03 kilograms for the low-GI group and an *increase* of 1.31 kilograms for the low-fat group.

Low Glycemic Index Diet And Weight Loss

Bouché C, et al. "Five-Week, Low–Glycemic Index Diet Decreases Total Fat Mass and Improves Plasma Lipid Profile in Moderately Overweight Nondiabetic Men." Diabetes Care, Volume 25, Pages 822-828. 2002. http://cat.inist.fr/?a Modele=afficheN&cpsidt=14158194

This report studied low-glycemic diets and high-glycemic diets. The average glycemic index of the low-glycemic diets was 38% for breakfast and 41% for lunch. The average glycemic index of the high-glycemic index diets was 75% for breakfast and 41% for lunch. The study showed that the low-glycemic diets lowered the levels of the postprandial (after meal) blood glucose and insulin much more than the high-glycemic diets did. The low-glycemic diets also reduced the fat body mass and increased the lean body mass. At the same time, the low-glycemic diets improved the blood lipid profile, which is linked to the risk of cardiovascular diseases. The study concluded, that low-glycemic diets could be beneficial in preventing metabolic disorders and cardiovascular complications.

High Glycemic Load And Coronary Heart Disease

Liu S, et al. "A prospective study of dietary glycemic load, carbohydrate intake, and risk of coronary heart disease in US women." American Journal of Clinical Nutrition, Volume 71, Number 6, Pages 1455-1461, June 2000. http://www.ajcn.org/cgi/content/full/71/6/1455

Foods with a low glycemic index are better for our health. However, we should also understand, the total amount of carbohydrates or glycemic load affects our health too. We should not assume that eating a large quantity of carbohydrates with a low glycemic index is safe. This study, from the "Nurse's Health Study," investigated the link between the consumption of carbohydrates and coronary heart disease.

The participants were grouped into five, based on the amount of daily carbohydrate consumption (glycemic load). A unit of glycemic load equals a gram of white bread or pure sugar. The first group, eating the lowest daily average amount of carbohydrates, had a glycemic load of 144 units, while the fifth group had 226 units. The risk of coronary heart disease for the fifth group was 1.98 times of that for the first group. That risk for the fifth group was nearly twice as high, yet these people ate only about 57% more in carbohydrates, or glycemic load, than the first group did. The study also showed a close relationship between a high glycemic load and the risks for coronary heart disease, especially in cases of women who had a BMI of 23 and higher.

A glycemic load of 144 units equals 144 grams of pure sugar. Keep it in mind that the first group of women in this study was not totally free of the risk of coronary heart disease. They ate the least daily carbohydrates and had the lowest risk of coronary heart disease in this study. To be free from the risk of coronary heart disease, the daily glycemic load must be lower than 144 units

High Glycemic Load Affects HDL And Triacylglycerol

Liu S, et al. "Dietary glycemic load assessed by food-frequency questionnaire in relation to plasma high-density-lipoprotein cholesterol and fasting plasma triacylglycerols in postmenopausal women." American Journal of Clinical Nutrition, Volume 73, Number 3, Pages 560-566, March 2001. http://www.ajcn.org/cgi/content/full/73/3/560?ijkey=2c7a9b 2d1321ba11a94133a4dad88d0879619f82

The study showed that the higher the dietary glycemic load, the lower the concentration of serum HDL. The lowest GL group of women with glycemic load at 117 units had HDL at 58 mg/dl. The highest GL group with glycemic load at 180 units had HDL at 52 mg/dl. ("dl" is deciliter; 100 ml or cc.) HDL is considered "good cholesterol," as related to its effect on coronary heart disease. The more we eat carbohydrates, the lower the HDL in our blood circulation and the higher the risk of coronary heart disease.

The study also found that the dietary glycemic load, especially the overall daily carbohydrate consumption was strongly and positively related to the concentration of fasting serum triacylglycerol. This was especially true of cases in which the BMI (Body Mass Index) of an individual was 25 and over. The highest GL group, with glycemic load at 180 units, had triacylglycerol at 155 mg/dl. The lowest GL group, with glycemic load at 117 units, had only 87 mg/dl. The study concluded that high glycemic load was a potential risk factor for coronary heart disease, especially in those who were prone to insulin resistance.

Dumping HDL With High Glycemic-Index Foods

Ford SF and Liu S. "Glycemic Index and Serum High-density Lipoprotein Cholesterol Concentration Among US Adults." Archives of Internal Medicine, Volume 161, Pages 572-576. 2001. http://archinte.ama-assn.org/cgi/content/full/161/4/572?ijkey=468df8e93cf846ab367b1ff36f76dded5a4995dc

The study found that the higher the glycemic index of the participant's food, the lower the concentration of the individual's HDL cholesterol. Comparing the groups rated from the lowest to highest glycemic index, the GI and HDL were 70.7% and 49.8 mg/dl for the lowest GL group, 77.4% and 45.4 mg/dl for the second, 81.3% and 43.8 mg/dl for the third, 85% and 42.4 mg/dl for the fourth, and 90.7% and 42.4 mg/dl for the highest group (Glycemic index for pure sugar or glucose is 100%.).

The study also observed that the above relationships were much stronger between HDL and the values of glycemic index when the results were adjusted by gender and BMI. Although the group who consumed the foods with the lowest glycemic index at 70.7% had a better blood lipids profile, we must realize that the group's glycemic index was still too high and unhealthy!

DIETARY RECOMMENDATIONS

Starchy Foods Are Linked To Diabetes

Ford SF and Liu S. "Glycemic Index and Serum High-density Lipoprotein Cholesterol Concentration Among US

Adults." <u>Archives of Internal Medicine</u>, Volume 161, Pages 572-576. 2001. http://archinte.ama-assn.org/cgi/content/full/161/4/572?ijkey=468df8e93cf846ab367b1ff36f76dded5a4995dc

This study found a link between eating high-glycemic-index foods, such as starchy white bread, and the risk of developing type 2 diabetes (DM). During a five-year study, the group that consumed foods with the lowest daily glycemic index (GI) at 20.8-46.0% (pure sugar or glucose as 100%) had a risk of diabetes (DM) at 1.00 (baseline for comparison); the second group, with a GI at 46.1-48.6% had 0.69 times the risks of diabetes (DM); the third group, with a GI of 48.7-51.5% had 1.08 times the risk; the highest group had a risk of 51.6-67.7% and 1.37 times the risk.

The study concluded that the carbohydrates with higher glycemic indices were positively related to the onset of diabetes (DM). The glycemic indices (in parenthesis) for the foods in the study were "white bread, rolls, or toast" (75), "crackers or crisp breads" (67), "rice, boiled (including brown rice)" (65), "cakes or sweet pastries" (64), "sweet biscuits " (63), "other breakfast cereal" (62), "whole-wheat or rye bread, rolls, or toast" (54), "fried rice" (48), "mixed dishes with rice" (48), "fruit bread" (48), "pasta or noodles" (48), "muesli" (46), "dim sims or spring rolls" (45), "pies or savory pastries" (45), "wheat germ" (41), "puddings" (39), and "pizza" (38).

Is it safe for us to eat those foods?

Decreased Feeling Of Hunger With Low Glycemic Index Diets

Ball SD, et al. "Prolongation of Satiety After Low

Versus Moderately High Glycemic Index Meals In Obese Adolescents." Journal of Pediatrics, Volume III, Number 3, Pages 488-494, in March 2003. http://pediatrics. aappublications.org/cgi/content/full/111/3/488

We believe that consuming more calories than our body needs can make us gain weight. We also assume that excessive fats in our diet can make us fatter. Most of us are still sure that eating more carbohydrates is the best for our health and keeps us slim. But the population is getting heavier as more people adopting the low-fat high-carbohydrate diets. Why?

This study observed the links between the feeling of hunger and the types of diets depending on their glycemic index GI). Theses diets were (1) low-GI meal replacement (LMR, GI=28%), (2) moderately high-GI meal replacement (HMR, GI=62%), and (3) low-GI whole food meal (LWM, GI=43%).

The study enrolled 16 obese adolescent participants, including 8 male and 8 female on three separate 24-hour admissions. The participants were randomly assigned to one of the three diets for breakfast and lunch on each of the three admissions. All of them completed all three diets during the study. The study showed that, after having a meal, the increase of blood glucose for the HMR group (GI=62) was at 100%, for the LWM group (GI=43) it was 57%, and for the LMR group (GI=28) it was 54%. The increase for insulin response to the increase of blood glucose for the HMR group (GI=62) was at 100%, for the LWM group it was 64%, and for the LMR group (GI=28) it was 49%. The time span after meal before a snack platter request was 3.1 hours for the HMR group and 3.9 hours for the LMR group. The study observed that the low GI diets had the smallest increase on blood glucose, the least insulin response; and a prolonged satiety after meal. The study also suggested that LMR was an effective diet for achieving a long-term weight control.

Health Benefits Of Low-Glycemic Diets

Pereira MA, et al. "Effects of a low-glycemic load diet on resting energy expenditure and heart disease risk factors during weight loss." Journal of American Medical Association. Volume 292, Number 20, Pages 2482-90. Nov 24, 2004. http://jama.ama-assn.org/cgi/content/ full/292/20/2482.

Cutting down the size of meals for weight loss is probably the most difficult job for many people, because the feeling of hunger is real. Hunger is not only physically but also physiologically. The feeling of empty stomach is usually tolerable to many people. However, the feeling of hunger, as a result of low blood sugar level (hypoglycemia), is intolerable to most of us, because of physical weakness and discomfort. Feeling ill with hunger is so discouraging that many people are ready to give up on restricting both calories and the amount of foods. The best dietary routine for weight loss is to keep the dieter satisfied with the diet, and to avoid the physiological hunger. The body continues to use energy. It reduces its use when resting. If the body spends more energy than it takes in, it will lose weight.

This study found that those with a low-glycemic load diet did not slow down the use of energy as much while resting as those with a low-fat diet did. The people with a low-glycemic load diet had less feeling of hunger than those with a low-fat diet. Improvement in insulin resistance, serum triglycerides, C-reactive protein (a protein produced by the liver reflecting the level of inflammation of the body), and blood pressure were superior in the people who maintained a low-glycemic load diet. The study concluded that a low glycemic load diet might help prevent or treat obesity, cardiovascular disorders, and diabetes mellitus.

Do Doctors Know About Dietary Recommendations?

Flynn M, Sciamanna, C, and Vigilante, K. "Inadequate physician knowledge of the effects of diet on blood lipids and lipoproteins." Nutrition Journal. Volume 2, Number 19. 2003. http://www.pubmedcentral.nih.gov/articlerender.fcgi?artid=305367

A doctor finishes up physical examination on an obese patient, and refills the prescriptions. The patient has some cardiovascular diseases. The doctor would advise the patient that he should watch his diet by staying away from fats and portioning his meals. Does it sound familiar? Sure! But, do doctors really know about dietary recommendation?

The National Cholesterol Education Program recommended dietary modification as the initial treatment for lowering cholesterols. In its third report, this program advocated lowering triglycerides as a secondary target to lowering LDL (Low Density Lipoprotein Cholesterol).

This study used a questionnaire to assess each physician's knowledge of basic nutritional physiology and the underlying recommendations in the Therapeutic Lifestyle Changes Diet for lower serum triglycerides. Triglycerides were a new target for the preventive treatment of coronary heart disease, at the time when this study was published. The result of the study showed a surprising number of physicians did not have correct or enough knowledge in nutrition. This report urged that to keep up with a growing need to modify patients' dietary lifestyle, physicians should become more knowledgeable about nutrition.

Rightly Educate Physicians And The Public About Nutrition

Makowske M and Feinman RD. "Nutrition education: a questionnaire for assessment and teaching." Nutrition Journal Volume 4, Number 2. 2005. http://www.pubmedcentral.nih.gov/articlerender.fcgi?artid=546238

Internists and family physicians are busy attending medical meetings for continuing medical education credits, but they rarely ever hear important nutrition topics that would aid them in their everyday practice.

This study was to assess the first-year medical students' knowledge of nutrition and metabolism.

The result showed physicians and the public, even the educated population, did not have correct knowledge in nutrition. For instance, while the third report of the National Cholesterol Education Program (ATP III) advocates lowering triglycerides as a secondary target to lowering LDL (Low Density Lipoprotein Cholesterol), the ATP III and other professional agencies still recommend low-fat (25-30%), high-carbohydrate (50–60%) diets. In fact, carbohydrates is the diet component responsible for increasing blood triglycerides, Authors of this study hoped that its results would help incorporate more nutrition study in the curriculum of medical education.

CARBOHYDRATES AND DIABETES

Most, if not all of us, have heard about diabetes (DM). Some people call it "sugar diabetes." In fact, there are two diseases with the name of diabetes. One is diabetes mellitus (DM), and the other is diabetes insipidus (DI). Although both diseases share a few similar symptoms such as ongoing thirst and frequent urination, they are two different, unrelated diseases. Please see Diabetes Insipidus and Diabetes Mellitus, Glossary of Important Terms, Chapter 1.

Carbohydrate-Rich Diets Cause Diabetes

"Prolonged exposure of human pancreatic islets to high glucose concentrations in vitro impairs the beta-cell function", which was published in the Journal of Clinical Investigation, 1992 October; 90(4): 1263–1268, by D L Eizirik, G S Korbuttbility, and C Hellerström of Department of Medical Cell Biology, Uppsala University, Sweden. http://www.pubmedcentral.gov/articlerender.fcgi?tool=pmcentrez&artid=443168

Theories about the causes of diabetes (DM) matter a good deal to those of us who would help prevent it from afflicting healthy men and women all over the world. We want to know how we may most effectively treat it, and perhaps even cure the disease for the diabetic men and women presently suffering from it. So far, the consensus is that diabetes (DM) is a genetic and/or a metabolic disorder. Rarely do many people question whether diabetes (DM) is a result of carbohydrate abuse. The human body is so wonderfully designed for adapting itself to a constantly changing environment. It has the capability of healing itself.

But, unlike the sky, it has a limit. It can be abused only so much before problems occur.

This study used healthy pancreatic ß-islet tissues (insulin-producing cells), which were harvested from a cadaver organ donor, to investigate the impact of glucose concentration on its development. The pancreatic ß-islet tissue was divided into three sets of specimens in the same regular culture solutions. (Please see Figure 35. Carbohydrate Abuse Causes Diabetes Mellitus.) In addition, one of the three sets of culture solutions contained either 5.6 mM (109.76 mg%), 11 mM (215.00 mg%) or 28 mM (548.80 mg%) glucose solution. The specimens were incubated for seven days. Culture solution or culture medium is a solution or preparation for growing tissue, bacteria, or virus. Our normal fasting blood sugar is between 70-100 mg%. A value of fasting blood sugar at 100-125 mg% and/or a value of blood sugar over 180 mg% after meal is considered pre-diabetic or diabetic. These values may be slightly different from one laboratory to the other. In this research, 5.6 mM (109.76 mg%) is about the normal fasting blood sugar level.

Immediately after incubation, none of the specimens showed changes in the DNA of the ß-islet cells or, "signs of morphological or structural damages." In comparing the amount of insulin secretion between all specimens and using the amount of insulin from the specimen with 5.6 mM (109.76 mg%) as 100%, the insulin secretion of the pancreatic ß-islet tissue was only 55% for the specimen with 11 mM (215.00 mg%) and 40% for the specimen with 28 mM (548.80 mg%), respectively during their incubation. Then all three ß-islet tissue specimens were exposed to a low glucose medium (1.7 mM or 33.32 mg%) for 60 minutes to "rejuvenate" the cells.

Next the specimens were exposed to a high glucose medium (16.7 mM or 327.32 mg%) for measurement of insulin secretion. The ß-islet tissue specimen, which was initially cultured at 5.6 mM (109.76 mg%), produced insulin at a higher rate when exposed to the low glucose medium (1.7 mM), and enhanced the rate of secretion when exposed to the high glucose medium (16.7 mM). However, both the ß-islet tissues specimens, which were initially cultured at 11 mM and 28 mM, respectively, did increase the rate of producing insulin when exposed to the low glucose medium (1.7 mM), but failed to enhance the rate when exposed to the high glucose medium (16.7 mM). The specimen, initially cultured at 28 mM, clearly showed a smaller production of insulin, compared to that at 5.6 mM. The cultures at 5.6 mM and 11 mM showed similar biochemical functions. However, the culture at 28 mM showed a decreased biochemical function. The results of this research imply that a healthy individual's pancreas can become diabetic if he overuses carbohydrates especially those with high glycemic indices. The results also question whether diabetes mellitus a genetic disease. Should we not determine whether we have abused our pancreas with too much carbohydrate foods?

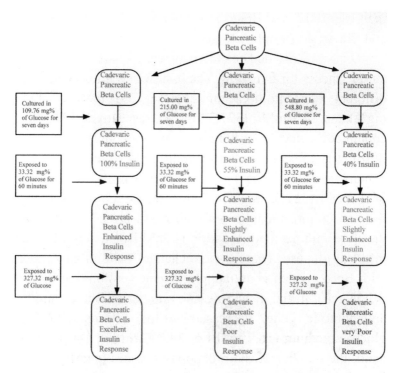

Figure 34. Carbohydrates Abuse Causes Diabetes Mellitus

More Pancreatic Cell Death In Hyperglycemia

Frederici M, et al. "High Glucose Causes Apoptosis in Cultured Human Pancreatic Islets of Langerhans." Diabetes. Volume 50, Number 6, Pages1290-1301, 2001. http:// diabetes.diabetesjournals.org/cgi/reprint/50/6/1290

Normal cells die after living for a normal span of time. *Apoptosis* means "death of the cell." Antiapoptosis means "prolonging the life of cells by outside influence such as genetic factors." Proapoptosis means "shortening the normal life by outside factors, again, such as genetic factors." The Bcl family genes are the factors in deciding the length of their life. Bcl-2 and Bcl-xl are antiapoptotic genes. Bad, Bid, and Bik are proapoptotic genes.

This study divided healthy pancreatic ß-islet cells in three culture media for five days. The first two solutions contained different concentrations of glucose, 5.6 mM/l (109.76 mg%) and 16.7 mM/l (327.32 mg%), respectively. The third contained 11 mM/l of mannitol (a sugar alcohol) and 5 mM/l of glucose for comparison or control. The number of dead pancreatic ß-islet cells was much higher in the culture with 16.7 mM/l (327.32 mg%) of glucose, as compared to both the first culture medium with 5.6 mM/l (109.76 mg%) and the third culture medium with 11 mM/l of mannitol and 5 mM/l of glucose. There was no change found in all cultures with Bcl-2, which is antiapoptotic gene (prolonging the life of the cell). In the meantime, the number of Bcl-xl, another antiapoptotic gene, decreased in the culture with high glucose medium (16.7 mM/l or 327.32 mg%). On the other hand, the number of the proapoptotic genes (shortening the life of the cell), Bad, Bid, and Bik, increased in the culture with high glucose medium (16.7 mM/l or 327.32 mg%). In conclusion, high blood glucose concentration could change the numbers of both the antiapoptotic and proapoptotic genes for causing death of the pancreatic ß-islet cells, and diabetes (DM).

Easing Diabetes With Low-Carbohydrate Diets

Boden G, et al. "Effect of a low-carbohydrate diet on appetite, blood glucose levels, and insulin resistance in obese patients with type 2 diabetes." Annals of Internal Medicine, Volume 142, Number 6, Pages 403-11. March 15, 2005. http://www.annals.org/cgi/content/abstract/142/6/403?etoc

Glycogenolysis is the conversion of glycogen in the liver and the muscles, into glucose. Gluconeogenesis is a renewal of glucose from protein and fat. Other than through glycogenolysis and gluconeogenesis, dietary carbohydrate is the main source of glucose for our body. Undoubtedly, if we take in less dietary carbohydrate, the amount of glucose in our body decreases. Diabetes (DM) is a disease that our body develops when it has more glucose that it can handle.

The study showed that for type 2 diabetic patients with a low-carbohydrate diet, both total daily calories and body weight decreased. Satisfaction with the diet increased while the feeling of hunger decreased. The study also found significant improvement (as related to the data before starting the diet) in lowering the average blood glucose to the normal range, dropping the reading of hemoglobin A_{1c} by 0.5%, increasing the insulin sensitivity by about 75%, reducing the average blood triglyceride by about one third, and the average blood cholesterol by one tenth. Please see Hemoglobin $A_{1c,}$ Glossary of Important Terms, Chapter 1.

Danger Of Eating Too Much Fructose

Basciano H, Federico L, and Adeli K. "Fructose, insulin resistance, and metabolic dyslipidemia." Nutrition & Metabolism, (Lond). 2005; Volume 2, Number 5, and online February 21. 2005. http://www.pubmedcentral.nih.gov/articlerender.fcgi?tool=pmcentrez&artid=552336

This is an Open Access article distributed under the terms of the Creative Commons Attribution License (http://creativecommons.org/licenses/by/2.0), which permits unrestricted use, distribution, and reproduction in any medium, provided the original work is properly cited.

Fructose is a simple sugar in fruit. It is also plentiful in dietary sources such as sucrose and corn syrup. This study cited that obesity and type 2 diabetes (the disease starts in the adulthood) had become an epidemic, in the United States and other parts of the world. Many studies showed a close link between the epidemic of obesity and type 2 diabetes from changes in the diet and lack of physical activities. One of the most important dietary changes was the increased consumption of fructose in sweeteners. After absorption, a large amount of fructose enters the liver, which is the key organ capable of breaking down fructose (catabolism). A high amount of glycerol and acyl group is the product of this breakdown, consequently resulting in dyslipidemia or high production of triglycerides and fat (lipogenesis). Dyslipidemia occurs when there is an abnormally higher amount of lipid in the blood circulation. The intestines also contribute to this high production. (See Figure 2 of this article below.) The article cited a 2004 study in which Gross and coworkers reviewed nutrient consumption in the US between 1909 and 1997, and found a strong correlation between the growing number of cases of diabetes (DM) and an increasing use of corn syrup. The study also found neither protein nor fat was responsible for the epidemic of diabetes (DM). The study urged that the public should be aware of the link between the high consumption of dietary fructose and the high incidence of insulin resistance and metabolic syndrome.

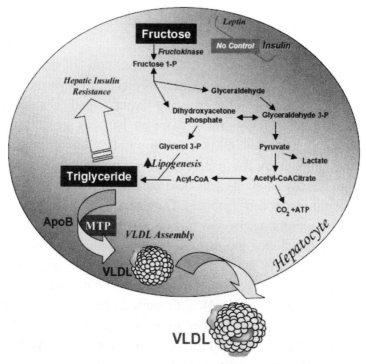

Figure 2
Hepatic fructose metabolism: A highly lipogenic pathway. Fructose is readily absorbed from the diet and rapidly metabolized principally in the liver. Fructose can provide carbon atoms for both the glycerol and the acyl portions of triglyceride. Fructose is thus a highly efficient inducer of *de novo* lipogenesis. High concentrations of fructose can serve as a relatively unregulated source of acetyl CoA. In contrast to glucose, dietary fructose does NOT stimulate insulin or leptin (which are both important regulators of energy intake and body adiposity). Stimulated triglyceride synthesis is likely to lead to hepatic accumulation of triglyceride, which has been shown to reduce hepatic insulin sensitivity, as well as increased formation of VLDL particles due to higher substrate availability, increased apoB stability, and higher MTP, the critical factor in VLDL assembly.

High Blood Sugar Causing Inflammation Of The Pancreas

Maedler K, et al. "Glucose-induced β cell production of IL-1β contributes to glucotoxicity in human pancreatic islets." Journal of Clinical Investigation, Volume 110,

Number 6, Pages 851–860. September 15, 2002. http://www. pubmedcentral.nih.gov/articlerender.fcgi?tool=pmcentrez&a rtid=151125

As of this date, most of us do not know all the causes for diabetes (DM). Classification of diabetes (DM) includes type 1, type 2, "other specific types," and gestational diabetes (DM). Gestational diabetes happens only during pregnancy. Diabetes (DM) is the result of failure of lowering blood glucose by the pancreas for whatever the reason(s). Type 1 diabetes (DM) was also called juvenile, childhood, or insulin-dependent diabetes. The disease most often begins in one's childhood. However, a small percentage of diabetic patients developed the disease as adults. The fact that type 2 diabetes (DM) causes chronic hyperglycemia (high blood sugar) is widely accepted in the medical community.

Type 1 diabetes (DM) is thought to be an autoimmune disease that destroys the insulin-producing ß-islet cells of the pancreas. IL-1ß (Interleukin-1ß) is a protein and cytokine, which promotes inflammation. It is plentifully found in cases of type 1 diabetes. Obviously, IL-1ß strains the ß-islet cells of the pancreas from producing insulin normally. Usually, a cell of the body dies (apoptosis) when Fas ligand or FasL, a protein, binds itself to its receptor on the cellular membrane (the capsule of a living cell). FasL belongs to the tumor necrosis factor (TNF or NF) family. In the case of type 1 diabetes (DM), IL-1ß also sets off the Fas-triggered cell death by activating NF-xB, a transcription (copying) factor for DNA.

This study reported that, after exposing the culture of ß-islet cells of the non-diabetic pancreas to high concentration of glucose, the concentration of IL-1ß markedly increased. Subsequently, the Fas-triggered cell death, activated by NF-xB, began. As a result of cell death, this study reported an increase in fragmented DNA, and destruction of the ß-islet cells. In summary as below:

High blood sugar →ß-cell of the pancreas → Producing IL-1β → Starting Fas-triggering cell death by activating NF-xB→ Death and destruction of cells with increase of DNA fragments.

Figure 35. Carbohydrates Abuse Kills the ß-Islet Cells
of the Pancreas

The study also reported an increase of IL-1ß -producing ß-islet cells in the pancreas of the type 2 diabetic patients, but not in the pancreas of the non-diabetic individuals. The study concluded that, other than possible genetic involvement, exposing the ß-islet cells of the non-diabetic person to chronic hyperglycemia could cause diabetes (DM). *(This study demonstrates that diabetes (DM) may not be a result of our metabolism disorders (metabolic syndrome) and is, at least, possibly a result of abusing carbohydrate foods.)*

Consuming Refined Carbohydrates Causes Diabetes

Gross LS, Li L, Ford ES, and Liu S. "Increased consumption of refined carbohydrates and the epidemic of type 2 diabetes in the United States: an ecologic assessment." American Journal of Clinical Nutrition, (1,2,3) Volume 79, Number 5, 774-779, May 2004. http://www.ajcn.org/cgi/content/full/79/5/774

Since the middle of last century, improvement in the quality of foods has continued, especially in refining carbohydrates. At the same time, type 2 diabetes in the US and world population has increased. This study found a significant correlation between type 2 diabetes and dietary fat, carbohydrate, protein, fiber, corn syrup, and total energy consumption. However, in evaluating each nutrient for total energy consumption, the study found a strong and positive correlation between type 2 diabetes and the consumption of corn syrup. The larger the consumption of corn syrup, the more new cases of diabetes (DM) were found. The study also found that as the consumption of fiber increased, the number of new diabetes (DM) cases went down. So long as there was no increase in the consumption of total energy, there was no correlation between type 2 diabetes and the consumption of fats.

High-Protein, Low-Carbohydrate Diet Helping Diabetes

Gannon MC and Nuttal FQ. "Effect of a High-Protein, Low-Carbohydrate Diet on Blood Glucose Control in People With Type 2 Diabetes." Diabetes, Volume 53, Number 9, Pages 2375-2382. 2004. http://diabetes.diabetesjournals.org/cgi/content/full/53/9/2375

There is an increasing number of weight loss programs currently being promoted. Also at present, there is a strong interest in formulating a dietary therapy for diabetics. This study used a diet with high-protein (30%), low-carbohydrate (20%), and (high-) fat (50%) to avoid weight loss and ketosis. The diet was "a low-biologically-available-glucose (LoBAG) diet". For comparison, the study used a regular diet with protein (15%), carbohydrate (55%), and fat (30%).

Ketosis happens when our body uses fat as the main

source of energy and produces ketones or ketone bodies. During starvation, the body must use fat to produce energy, and produces ketones. In observation, hours of starvation can lower the blood sugar in type 2 diabetes (DM). The study found the average 24-hour blood sugar was much lower with the LoBAG diet than the regular diet, 126 mg% versus 198 mg%. Besides, the blood sugar continued to fall, even after discontinuing the LoBAG diet. The amount of insulin in the circulation was also decreased with the LoBAG diet. The study found a significant improvement in the percentage point of glycosylated hemoglobin or hemoglobin A_{1c} in the untreated type 2 diabetic participants who were on the LoBAG diet. The study suggested that the LoBAG diet could possibly improve the untreated type 2 diabetic patients' state of high blood sugar, by adapting themselves without the use of medications. (Please see the figure below.) The blue (test group) and yellow (control group) triangles were the values of 24-hour blood glucose. The blue (test group) circles were the lowered values of the 24-hour blood glucose after five-week of the LoBAG diet, and the yellow (control group) circles were the values of 24-hour blood glucose after five-week of the regular diet.

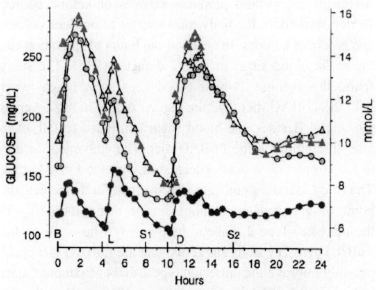

Effect of diet on glucose. Mean plasma glucose concentration before (triangles) and after 5 weeks on control diet (yellow circles: (CHO: fat: protein = 55:30:15)) or 5 weeks on the higher fat diet (blue circles: (20:50:30)). Meals are Breakfast (B), lunch (L) and dinner(D) plus 2 snacks (S1, S2). Data from this article, "Effect of a High-Protein, Low-Carbohydrate Diet on Blood Glucose Control in People With Type 2 Diabetes", modified for the article, "When is a high fat diet not a high fat diet?", that was published in the Nutrition & Metabolism (London). Online 2005 October 17. doi: 10.1186/1743-7075-2-27, by Richard D Feinman
Reprint with permission from Professor Richard Feinman

Normal Weight With Bad Health

Ruderman N, et al. "The metabolically obese, normal-weight individual revisited." Diabetes, Volume 47, Number 5, Pages 699-713. May 1988. http://diabetes.diabetesjournals.org/cgi/reprint/47/5/699

All of us probably know someone who has never been overweight or obese, despite the fact that he eats everything without restriction. Then we learned that this person has become diabetic (DM) or suffered a heart problem, stroke, arthritis, or cancer. How could it happen?

Since 1980's we have paid a growing attention to the young and non-obese with BMI (Body Mass Index) between 20 and 27 (Overweight: BMI between 25 and 29.9). Study reports showed that this apparently healthy segment of the population could suffer hyperinsulinemia (increase of insulin in the blood circulation), insulin resistance (the tissue does not use glucose despite the increase of insulin in the blood circulation), central fat distribution (more fat in the belly or trunk), type 2 diabetes, and premature coronary heart disease. This study reaffirmed that, young adults who gained only 2-10 kilograms (4.5 pounds to 22 pounds) of body weight in their young adulthood, like the obese population, had a potential of suffering hyperinsulinemia, insulin resistance, and central obesity of suffering diabetes and cardiovascular diseases. However, this study observed that these young, non-obese patients would have a better chance to reverse and avoid all the ill consequence in their later life, if they began to change their diets, to exercise regularly, and to start medications early on if indicated.

A Dangerous Sign With Fat In The Belly

Freedland ES. "Role of a critical visceral adipose tissue threshold (CVATT) in metabolic syndrome: implications for controlling dietary carbohydrates: a review." Nutrition & Metabolism, (Lond). Volume Number 12. 2004. Published online November 5. 2004. http://www.pubmedcentral.nih. gov/articlerender.fcgi?tool=pmcentrez&artid=535537

More evidence shows that increasing the amount of visceral adipose tissue is closely linked to hyperinsulinemia and insulin resistance. Hyperinsulinemia and insulin resistance lead the path to becoming diabetes and "metabolic syndrome". Viscera mean internal organs particularly those in the belly. Adipose means fat.

There are two types of "fat in the belly". One is accumulation of fat underneath the skin or the subcutaneous layer. This type of fat accumulation is usually not as risky as the other type, in terms of its link to diabetes (DM), insulin resistance, hyperinsulinemia, and cardiovascular disease. The other type of fat accumulation is underneath the abdominal wall and inside the abdominal cavity. Fat is deposited around the internal organs such as bowels and stomach. Also, fat is deposited behind the abdominal cavity.

This study suggested that each individual has a "critical threshold" for the increase of the visceral adipose tissue, before becoming symptomatic with the metabolic syndrome. The individualized threshold would explain why some non-obese individuals suffer metabolic syndrome while some obese people would not. The study also suggested that controlling dietary carbohydrates might help prevent and treat the metabolic syndrome.

Reduce Blood Sugar (Carbohydrates) For Health

Stratton IM, el al. "Association of glycaemia with macrovascular and microvascular complications of type 2 diabetes (UKPDS 35): prospective observational study." British Medical Journal (BMJ). Volume 321, Number 7258, Pages 405–412. August 12, 2000. http://www.pubmedcentral. nih.gov/articlerender.fcgi?tool=pmcentrez&artid=27454

Hemoglobin A_{1c} (HbA_{1c}) indicates the amount of globin (protein) out of total hemoglobin being bonded with glucose in our body. HbA_{1c} is expressed in percentage point. It is an indicator for estimating the level of blood sugar and the progress of diabetes. The higher the amount of glucose in our body, the higher the percentage of HbA_{1c} will be. The normal HbA_{1c} is below 6%.

This study indicated that, by dieting, the lowering of 1% in HbA_{1c} (or, 35 mg% of blood glucose) would reduce the risks of various clinical disorders as follows: 21% for diabetes; 14% for cardiovascular disease; and 37% for microvascular (small blood vessels) diseases.

Dietary Fat And Risks Of Cardiovascular Disease

He K, et al. "Dietary fat intake and risk of stroke in male US healthcare professionals: 14 year prospective cohort study." British Medical Journal. Volume 327, Number 7418, Pages 777–782. October 4, 2003. http://www.pubmedcentral. gov/articlerender.fcgi?tool=pmcentrez&artid=214078

Fat has been a "bad" or dirty" word. No one would consciously praise "fat" in any form or shape under any circumstance. For more than half a century public health agencies, both at the world and national levels, have actively promoted high-carbohydrate low-fat diets. Their campaign has been based on the seriously incorrect notion that dietary fat is making the body fat. Moreover, the fat and cholesterol in our blood circulation have been blamed for many diseases such as heart attack, stroke, diabetes (DM), cancers, and others. We have become afraid of fat or "lipophobic." But, do we really need to be lipophobic?

This 14-year study investigated the link between an

individual's risk of stroke and the total amount of fats (including animal and vegetable fats) they ate daily. The study found that the risk of stroke for a group with the lowest daily consumption of total fats at 54 grams was 1.0. The risk of stroke for a group with the highest daily consumption of fats at 86 grams was 0.91. The study concluded that there was no evidence to link the intake of fat, types of fat, or cholesterol to the risk of stroke in men.

Eggs And Risks Of Coronary Heart Disease

Kritchevsky SB and Kritchevsky D. "Egg consumption and coronary heart disease: an epidemiologic overview. " Journal of the American College of Nutrition. Volume 19, Number 5 (Supplement), Pages 549S-555S. October 2000. http://www.ncbi.nlm.nih.gov/entrez/query.fcgi?cmd=Retrieve&db=PubMed&list_uids=11023006&dopt=Citation

We know that the egg yolk contains plenty of cholesterol, thanks to public health agencies, physicians, and nutritionists. They told us to not eat egg often, because of the cholesterol in the egg. But, how true is it that eating egg increases cholesterol in our blood? A few researches have investigated this issue, and found that the message in the campaign, at least, incorrect.

This study reviewed the epidemiologic data from different studies to find the link between the consumption of eggs and the risk of coronary heat disease. Dietary cholesterol intake might be slightly linked to the risk of coronary heart disease. When considering a complete range of co-factors the risk of coronary heart disease increased about 6%, for eating each 200 mg of cholesterol, per 1000 kilocalories of food. The study concluded that there was no apparent evidence that suggested a link between the consumption of more than one egg a day and the risk of coronary heart disease in non-diabetic men and women. A recent study showed that eating eggs could help increase HDL-c and decrease inflammatory factors.

Low Carbohydrate Diet For Diabetes Mellitus

Arora SK and McFarlane SI. "The case for low carbohydrate diets in diabetes management." Nutrition & Metabolism, (Lond), 2005; Volume 2, Number 16. 2005. Published online 2005 July 14. http://www.pubmedcentral. nih.gov/articlerender.fcgi?tool=pmcentrez&artid=1188071

While we emphasize prevention and early treatment for diabetes (DM), we had not paid much attention to the effectiveness of ongoing dietary changes in treating the disease. We use so many kinds of medications everyday then talk about the possible complications. We have had little success in stopping the progress of diabetes (DM). We should realize that effective treatment must try to eliminate the original cause of the disease. In the case of diabetes (DM), it means carbohydrates in the diet. This study reviewed 88-plus research articles in analyzing the relationship between the promotion of the low-fat, high-carbohydrate diets during the past three decades and the epidemic of diabetes mellitus and its complications. The study explained the failure in treating diabetes with a combination of a low fat, high carbohydrate diet, and exercise. The study also agreed that the high carbohydrate diet should be faulted for an increasing rate of obesity. The diet should also be blamed for the risks of diabetes mellitus, hypertension, abnormal blood lipids, and cardiovascular disease.

The study also cited a close relationship between the decrease in fat consumption and the increase of obesity and diabetes (DM) over the last three decades. It observed that some forms of carbohydrate restriction showed weight reduction and positive change in the symptoms of diabetes mellitus, the values of blood lipids, and other benchmarks for the risks of cardiovascular disease. The study cautioned against the suggestion of eating less than 30 grams of carbohydrates everyday. In the meantime, the study cited no scientific basis for objecting to the low carbohydrate diet.

Another Low Carbohydrate Diet For Diabetes Mellitus

Nielsen JV and Joensson E. "Low-carbohydrate diet in type 2 diabetes. Stable improvement of bodyweight and glycemic control during 22 months follow-up." Nutrition & Metabolism, (Lond), Volume 3, Number 22. 2006 Published online June 14, 2006. http://www.nutritionandmetabolism. com/content/3/1/22

During recent years more researchers are looking into the link between high-carbohydrate diets and diabetes (DM). When we can establish the relationship between the cause and the outcome of a disease, we should be able to treat the disease logically. That means that if we can remove the cause of the disease, we may find a cure for it. Inasmuch as several studies have shown that high blood sugar can cause diabetes (DM), there is a hope that restricting carbohydrates in the diabetic patient's diet can possibly cure, or at least ease up, his diabetic conditions.

This Swedish study previously reported that a 20% carbohydrate diet was much better than a 55-60%

carbohydrate diet for helping obese diabetic patients control their blood glucose level and lose weight for a period of six months.

Twenty-two months after starting the first study, this study further investigated the possible therapeutic effects with a low-carbohydrate diet for diabetic patients. All but one of the 16 patients had lower body weight at the end of 22 months. In review, the average HbA_{1c} dropped from 8.0 ± 1.5% to 6.9 ± 1.1%, in 22 months. This demonstrated that the 20% carbohydrate or a low-carbohydrate diet, with some calorie restriction could help treat diabetes (DM).

After this study was published, the leadership of a medical professional organization still rejected the call. He insisted that low carbohydrate diets were too restrictive, that they were difficult for diabetic patients to continue for a long period of time. Dieting is a personal choice. A diet with more fat and protein can provide the dieter better satiety than a diet with abundant carbohydrates. The latter often makes the dieter eat more, and want to eat more, because of a vicious cycle, or that I named *Sweet Roller Coaster,* between eating more carbohydrates, rising blood glucose level, more secretion of insulin, lowering blood sugar, feeling hunger, and the urge to eat more carbohydrates. (Please see Figure 24. Sweet Roller Coaster.) Trying to hold off the vicious cycle with low fat, high carbohydrate diets has proved to be a failure!

CARBOHYDRATES AND CHOLESTEROL

Blood or plasma lipoproteins are molecular complexes of lipid (lip- or fat) and protein. (Please see Figure 18 Types of Plasma Lipoproteins, Glossary of Important Terms,

Chapter 1.) Different lipoproteins have different molecular weights and shapes. Apolipoprotein is the protein part of a lipoprotein. Apolipoprotein A is the protein part of high density lipoprotein (HDL); apolipoprotein B is that of low density lipoprotein (LDL); and apolipoprotein C is that of intermediate density lipoprotein (IDL) and very low density lipoprotein (VLDL). Either cholesterol or triglycerol is the lipid part of lipoproteins. Most commonly known lipoproteins are enzymes, toxins, adhesins (substance involved in adhesion), antigens (protein stimulates one's immune system for producing antibody and causing antigen-antibody reaction), and transporters. The availability and amount of both the protein and lipid components and the rate of reaction between both components determine the amount of the lipoproteins. We commonly express the complex of apolipoproteins and cholesterol as HDL-c, LDL-c, IDL-c, and VLDL-c.

The concentration or amount of both HDL-c and LDL-c, especially VLDL-c and their ratio, are the indicators for the risks of cardiovascular diseases. HDL-c is considered good lipoprotein-cholesterol, because HDL-c has smaller amount of cholesterol (and triglycerides) tightly held inside the protein layer. On the other hand, LDL-c is considered bad lipoprotein-cholesterol, because LDL-c has much more of cholesterol (and triglycerides) loosely held inside the protein layer. As a result of the structure, LDL-c is much easily release cholesterol to the wall of the inflamed artery. Among all subgroups of the LDL-c, VLDL-c is really the bad one. In the presence of high blood glucose, which causes inflammation of the blood vessels, VLDL-c attacks and deposits its cholesterol onto the inner layer of the wall of the blood vessels, especially the arteries.

Triglycerides are molecules made up of glycerol (three alcohol chains) bonded with three molecules of fatty acids. Glucose is the material for producing cholesterol. (Please see Figure 5 Metabolism of the Nutrients and Figure 16 Insulin and Glucose Metabolism, Chapter 1.) A high blood glucose level produces more cholesterol and VLDL-c. Except under the situations of low blood glucose, all the blood sugar is coming from dietary carbohydrates. Therefore, eating too many carbohydrates causes a high blood glucose and high cholesterol and triglyceride levels.

Carbohydrate-Rich Diets, High Triglycerides And Fat

Schwarz JM, Linfoot P, Dare D, and Aghajanian K. "Hepatic de novo lipogenesis in normoinsulinemic and hyperinsulinemic subjects consuming high-fat, low-carbohydrate and low-fat, high-carbohydrate isoenergetic diets." American Journal of Clinical Nutrition, Volume 77, Number 1, 43-50, January 2003. http://www.ajcn.org/cgi/content/abstract/77/1/43

For the recent couple of decades, public awareness of the risks of coronary heart disease has increased. Fats, especially cholesterol, in the blood circulation have become a hot household topic. In the meantime medical professionals, following the trend, monitor their patient's blood tests for cholesterol and lipid, and prescribe more Statin drugs to lower their patient's cholesterol and LDL or LDL-c. There are more studies being done to find out the relationship between diets and the production of cholesterol, especially VLDL-c. VLDL-c is a subgroup of LDL-c, and really the worst LDL-cholesterol for arteriosclerosis and a risk for coronary heart disease. For the past few years, triglycerides have also been identified as an independent risk factor, only second to LDL-c as a danger for coronary heart disease. The liver produces fatty acids, triglycerols (triglycerides), and cholesterol with glucose. Undisputedly, through understanding the relationship between diets and the production of cholesterol, we can alter one to improve the other.

This study investigated the link between a low-fat, high-carbohydrate diet and the production of the new fatty acid (de novo lipogenesis or DNL) and triglycerides in the blood circulation by the liver. It found the high-fat, low-carbohydrate diet increased new free fatty acids (DNL). The increase was 3.7-5.3 times higher in the patients who were obese and had a high blood level of insulin than those who were either lean or obese but had a normal blood level of insulin. It also found that the high-fat, low-carbohydrate diet did not change the VLDL-triglyceride concentration. The study also showed that both a high blood level of insulin and following a low-fat high-carbohydrate diet could increase the production of fatty acids by the liver. At the same time, the production of new fatty acids by the liver was linked to a high concentration of triglycerides in the circulation, as a result of production by the liver with glucose. (Please see Figure 17 Insulin and Cellular Utilization of Glucose, Glossary of important terms, Chapter 1.)

With the understanding of the relationships among high blood level of insulin (hyperinsulinemia), high-carbohydrate diets, high blood sugar in diabetes (DM), and the concentration of VLDL-triglycerides, we should be more careful in promoting high carbohydrate diets to both the diabetic patients and the general population.

Men's Lipids Improved With Very Low Carbohydrate Diet

Volek JS, Gómez AL and Kraemer WJ. "Fasting Lipoprotein and Postprandial Triacylglycerol Responses to a Low-Carbohydrate Diet Supplemented with n-3 Fatty Acids." Journal of the American College of Nutrition, Volume 19, Number 3, 383-391 2000. http://www.jacn.org/cgi/content/abstract/19/3/383?ijkey=0a67e87126edc558acd61b8e38d9028199d4f601&keytype2=tf_ipsecsha

Since low-carbohydrate diets became popular in the United States, their safety and long-term impacts on the dieter's health are still uncertain. The most important issues are blood lipids, coronary heart disease, and diabetes (DM), among others. This study found that, with a very low carbohydrate (5-10% of the daily calories) high fat diet, fasting and postprandial (after meal) triglycerides were significantly lowered. The blood levels of cholesterol and LDL were temporarily increased. The study cited the high fat very-low carbohydrate diet was "a clinically significant positive adaptation in terms of cardiovascular status." However, they believe more studies should be conducted to find out which subgroup of LDL-cholesterol was increased with the diet.

Improve Men's CVD Biomarkers With Ketogenic Diet

Sharman MJ, et al. "A Ketogenic Diet Favorably Affects Serum Biomarkers for Cardiovascular Disease in Normal-

Weight Men." Journal of Nutrition Volume 132, Number 7, Pages 1879-1885, 2002. http://jn.nutrition.org/cgi/content/ab stract/132/7/1879?ijkey=66664486

The focus of this study was the outcome of a ketogenic diet that included 30% of daily calories in proteins, 8% in carbohydrates, 61% in fats. This study found that the ketogenic diet caused 33% less fasting serum triglycerol, 29% less for postprandial lipids, and 34% less for the postprandial insulin concentration, than the regular diet. The ketogenic diet caused no increase in the concentration of total cholesterol, LDL cholesterol, and oxidized LDL. The result of this study suggested a short-term ketogenic diet might improve the risk of cardiovascular disease.

Women's Lipids Improved With Very Low Carbohydrate Diet

Volek JS, et al. "An Isoenergetic Very Low Carbohydrate Diet Improves Serum HDL Cholesterol and Triacylglycerol Concentrations, the Total Cholesterol to HDL Cholesterol Ratio and Postprandial Lipemic Responses Compared with a Low Fat Diet in Normal Weight, Normolipidemic Women." Journal of Nutrition Volume 133, Number 9, Pages 2756-2761, September 2003. http://jn.nutrition.org/cgi/content/ abstract/133/9/2756

This study investigated the impact on blood lipids of women with an isoenergetic (same amount of calories) very-low-carbohydrate diet, in comparison with a regular diet. The study observed that the very-low-carbohydrate diet resulted in an increase of the fasting serum cholesterol by16%, the LDL-cholesterol by15%, and the HDL-cholesterol by 33%. At the same time, it caused a decrease in triacylglycerols by 30%, the ratio of total cholesterol to HDL- cholesterol by 13%, the 8-hour postprandial triacylglycerol by 31%. The study concluded that a very-low carbohydrate diet had a favorable short-term impact on the risk of cardiovascular diseases for healthy women.

The Carbohydrate-Rich Diet Increases VLDL

Hudgins LC, et al. "Human Fatty Acid Is Stimulated by a Eucaloric Low Fat, High Carbohydrate Diet." Journal of Clinical Investigations, Volume 97, Number 9, Pages 2081–2091. May 1996. http://www.pubmedcentral.nih.gov/picrender.fcgi?artid=507283&blobtype=pdf

If you are familiar with organic chemistry, you may know the molecules mentioned in this study. If you are not, just remember the names and their important roles in our health. Palmitic acid is a saturated fatty acid, consisting of 16 carbon atoms, produced in lipogenesis, which is a synthesis of fatty acid in the liver from other chemical substances. Linoleic acid is an 18-carbon polyunsaturated fatty acid. *Polyunsaturated* represents multiple sites of bonding, which are unsaturated between carbon atoms. Linoleic acid is an essential fatty acid that cannot be produced in the body of all mammals.

Because of concerns over the increase of cholesterol in the blood circulation and body fat, medical and nutritional experts have urged people to eat more carbohydrates with as little fat and cholesterol as possible. They thought that the dietary fat and cholesterol are responsible for the increase of cholesterol in the blood circulation and body fat. High cholesterol in the blood circulation is the risk for cardiovascular diseases.

This study investigated the changes in the amount of fatty acid produced by the body, after eating either a carbohydrate-rich diet or a fat-rich diet for 25 days. The carbohydrate-rich diet had 75% of calories from carbohydrates and 10% from fats. The fat-rich diet had 45% of calories from carbohydrates and 40% from fats. The diets contained same amount of calories.

The study showed that the carbohydrate-rich diet caused a significant increase in the amount of VLDL triglycerides, which were plentiful with palmitate and scarce with linoleate. The study also found that about 44% of the VLDL triglycerides was newly synthesized. In contrast, all the individuals on the fat-rich diet showed an insignificant change in VLDL synthesis. The study suggested that preferring carbohydrates to fats in diet might cause adverse effects on the cardiovascular system.

CARBOHYDRATES AND INFLAMMATION

Over the past half a century, medicine seems having made so much advancement. It has discovered new diseases and, in many cases, found the causes of those diseases. It studied how the diseases start and develop, even at the cellular and molecular levels. It made new drugs to treat diseases, to intercept or prevent diseases, and to ease or alleviate the symptoms. It developed techniques and instruments for repairing our diseased body.

We are indeed grateful for many wonderful accomplishments in today's medicine. We have seen more people living to old age, while others are dependent upon medications and surgeries. Many people in the latter group may still be working and productive, yet they are disabled to some extent.

We claim that we have discovered the causes of diseases and then found solutions to preventing those causes. At the same time, we seem unable to find definite causes for many other diseases. Apparently, there is still a piece of missing puzzle in today's medicine.

More and more we have learned that inflammation is the cause of a growing list of diseases. The most common diseases are inflammatory diseases such as arthritis and fibromyositis (or fibromyalgia). We have developed many anti-inflammatory drugs in hope of treating those diseases. But we have not and cannot! What is the problem that prevents us from developing a successful treatment? The answer is that we still do not get to the real cause of the diseases, that we are treating the diseases only after they have already started.

We must understand two things. One is the reason why the body gets inflamed. The other is which of our daily nutrients (carbohydrates, fats, and proteins) may contribute to promoting inflammation in the body. In the process of inflammation, there are a few names we need to know. These names are useful in understanding the process of inflammation itself. If you have already known them, that is great. If you do not know them, just pick up their names and understand a little bit of what they are doing.

White blood cells or leukocytes are cells of our immune system. White blood cells include neutrophils; basophils; eosinophils; lymphocytes; monocytes, and macrophages, White blood cells defend our body against invasion by bacteria, viruses, parasites, and foreign bodies.

- Neutrophils are the main elements in acute

inflammation, such as a result of infection by bacteria and fungi. They have the ability of phagocytosis that gobbles the bacteria, fungi, and foreign particles. Pus is a result of their activities and death.

- Basophils are primarily responsible for allergic and antigen reactions; they release histamine, which causes inflammation.
- Eosinophils deal with infection by parasites.
- Lymphocytes are found mostly throughout the lymphatic system. There are three types of lymphocytes in the blood circulation. They produce antibodies; kill bacteria inside cells and viruses, and kill infected and tumor cells.
- Monocytes have the ability of phagocytosis, like neutrophils.
- Macrophages are those monocytes, which develop into specializing in phagocytosis when they migrate into the tissues from the blood stream.

There are a few names of molecules often involved in inflammation, such as cytokines, interleukins-6, and C-reactive protein (CRP).

- Cytokines are proteins or a group of peptides. Cytokines play an important role in our immune responses, particularly in many kinds of immunological reactions, inflammation, and infections.
- Interleukins-6 (IL-6) are cytokines, that promote inflammation.
- T-cells are a kind of lymphocyte circulating in the blood stream.
- Macrophages (also listed above) are a kind of

monocytes, which are specialized in phagocytosis in the tissue.

• Endothelial cells are the cells of the inner layer of the wall of the blood vessel. In the case of acute inflammation, T-cells, macrophages, and the endothelial cells produce IL-6. Production and secretion of IL-6 trigger the immune system in response to tissue damage.

• C-reactive protein (CRP) is produced by the liver, in response to the level of interleukins-6 of the blood circulation. C-reactive protein is thought to bind itself to damaged cells or foreign bodies, to facilitate phagocytosis by macrophages. The level of CRP is an indicator for the level of inflammation of our body.

Carbohydrate Diet And C-Reactive Protein

Seshadri P, et al. "A randomized study comparing the effects of a low-carbohydrate diet and a conventional diet on lipoprotein subfractions and C-reactive protein levels in patients with severe obesity." American Journal of Medicine, Volume 117, Issue 6, Pages 398-405. September 15, 2004. http://www.ncbi.nlm.nih.gov/sites/entrez?db=PubMed&cmd=Retrieve&list_uids=15380496

Inflammation has gained attention in recent years. We have come to understand that many more diseases are a result of inflammation, including diabetes (DM), coronary heart disease, arthritis, fibromyositis, dementia and Alzheimer's disease, and cancer. If we know the real cause of inflammation, we can avoid or remove the cause to prevent inflammation from happening in the first place.

This study involved a low-carbohydrate diet, and a

conventional diet with restrictions on fat and calories for the investigation of the link between diets and inflammation. The study found the low-carbohydrate diet resulted in much lower blood concentration of VLDL than those on a conventional (high-carbohydrate, low-fat, and low-calorie) diet. However, the low-carbohydrate diet increased chylomicrons in the blood circulation. Chylomicron are a lipoprotein formed in the intestine that absorb and transport triacylglycerol, cholesterol, and fat-soluble vitamins. Both diet groups had a decrease in the concentration of LDL (low-density lipoprotein) and an increase in HDL (high-density lipoprotein).

The study also observed the low-carbohydrate diet decreased the concentration of C-reactive protein much more than the conventional diet did. This study concluded that a low-carbohydrate diet offers a positive effect on the high-risk, obese patients, with improvement in both the subgroup of lipoprotein (VLDL) and the level of inflammation marker (CRP).

High Blood Sugar Reduces Resistance To Infection

Van Oss CJ. "Influence of Glucose Levels on the In Vitro Phagocytosis of Bacteria by Human Neutrophils." Infection and Immunity, Volume 4, Number 1, Pages 54–59. July 1971. http://www.pubmedcentral.nih.gov/articlerender.fcgi?tool=pmcentrez&artid=416264

When phagocytosis is weakened by any reason, the body will suffer a prolonged infection and inflammation. This study observed that phagocytosis of the neutrophils was most active when the concentration of a person's blood sugar was at the level between 50 mg% and 100 mg%. As the level of blood sugar went up to 200 mg%, 400 mg%, and 800 mg%, the activity of phagocytosis was increasingly suppressed. This is the reason why diabetic patients tend to suffer prolonged bacterial infections that are resistant to treatment. The most important step toward preventing infection and inflammation may be taken by controlling our blood sugar level. The best way to control our blood sugar level is to properly cut down the amount of carbohydrates of our diet.

Hyperglycemia And Failed Organ Transplants

Thomas MC, et al. "Early peri-operative hyperglycaemia and renal allograft rejection in patients without diabetes." Nephrology (BioMed Central). Volume 1, Number 1. 2000. Published online October 4, 2000. http://www.pubmedcentral. nih.gov/articlerender.fcgi?tool=pmcentrez&artid=29098

Even when surgeons make a compatible allograft (an organ transplant from a donor to a recipient of the same specie, such as a man to a man), they may witness its rejection by the recipient's body. The observation has been particularly true in cases involving diabetic recipients.

This study investigated the link between the early-on rejection rates in the compatible, allograft kidney from cadaver and the blood sugar levels of the organ recipients. The study cited an established finding that an acute high blood sugar level can increase (1) injuries by lack of blood supply; (2) presence of antigen to cause antigen-antibody reaction; (3) cell death (apoptosis); and (4) escalation of inflammation. The study found that the rejection rate was 42% in non-diabetic patients whose blood sugars after surgery were below 153.6 mg%. However, the rejection rate sharply increased to 71% of the non-diabetic patients whose blood sugars after surgery were over 153.6 mg%. A similar result was observed in the diabetic patients.

C-Reactive Protein And Risks Of Cardiovascular Diseases

Yudkin JS, Stehouwer CDA, Emeis JJ and Coppack SW. "C-Reactive Protein in Healthy Subjects: Associations With Obesity, Insulin Resistance, and Endothelial Dysfunction. A Potential Role for Cytokines Originating From Adipose Tissue." Arteriosclerosis, Thrombosis, and Vascular Biology, (the American Heart Association, Inc.) Volume 19, Number 4, Pages 972-978. 1999. http://atvb.ahajournals.org/cgi/content/full/19/4/972? ijkey=bc7ff94a637764853fbee0ded041fcffb04b0cff

In an acute inflammatory state, the liver produces C-reactive protein in response to the level of inerleukin-6 (IL-6), which the subcutaneous fat tissue produces. Interlukin-6 is a pro-inflammatory cytokine. We use the level of C-reactive protein as a risk factor of coronary heart disease.

Blood concentrations of the von Willebrand factor, tissue plasminogen activator, and cellular fibronectin, are the indicators for the functions of the endothelium (the inner layer of the blood vessel). The von Willebrand factor, tissue plasminogen activator, tissue plasminogen, and cellular fibronectin all involve blood clotting. They are also involved in tissue healing with collagen, heparin, and fibrin. Collagen is protein and an important component of connective tissue of animals, such as vascular wall and skin. Heparin is an anticoagulant or a molecule for preventing blood clotting.

Fibrin is a protein in clotting blood; it builds up a mesh to hold the platelets for building a blood clot.

This study reported the levels of C-reactive protein increased as the blood concentration of interlukines-6 (a pro-inflammatory cytokine) and tumor necrosis factor--α (a factor secreted by the adipose or fatty tissue) increased in cases of obesity and chronic infection. The study also observed a positive link of C-reactive protein to both insulin resistance and endothelial damage. Endothelial damage is present with increases in the blood concentration of the von Willebrand factor, tissue plasminogen activator, and cellular fibronectin. Insulin resistance is a state of high blood glucose in spite of a high concentration of insulin in the blood circulation.

The study observed that (1) a low–grade, chronic inflammation might cause insulin resistance and endothelial damage, (2) insulin resistance was leading to obesity, and (3) endothelial damage increased the risk of cardiovascular disease. *But, how about the other way around? Perhaps high blood sugar causes hyperinsulinemia and chronic inflammation, in which the concentration of IL-6, TNF-α, and CRP increase.*

Hemoglobin A$_{1c}$ And Fibromyalgia

Tishler M, et al. "Fibromyalgia in diabetes mellitus." Rheumatology International, Volume 23, Number 4, Pages 171-3. July 2003. Electronically published 2003 May 20. www.ncbi.nlm.nih.gov/sites/entrez?cmd=Retrieve&db=pubmed&dopt=AbstractPlus&list_uids=12756495&itool=iconabstr

Hemoglobin is the component of the red blood cell

and carries oxygen. Hemoglobin consists of heme, which contains iron, and globin. Globin is a protein that can bond itself with the blood glucose in the process of glycosylation. Hemoglobin A_{1c} is the portion of the total hemoglobin expressed in percentage points of which the globin is bonded with glucose. Hemoglobin A_{1c} is used as an indicator for the level of blood glucose; because the higher the blood glucose level is, the higher the Hemoglobin A_{1c} will be. It also reflects the individual's potential of becoming diabetic and developing other health complications as a result of high blood glucose.

This study found that 17% of the diabetic patients and only 2% of the non-diabetic group were positive for fibromyalgia. There was no significant difference in susceptibility of fibromyalgia between the groups of type 1 and type 2 diabetic patients. The diabetic patients who suffered more serious fibromyalgia had higher percentage points of HbA_{1c}. This study suggested that improved control of blood glucose levels might prevent fibromyalgia.

Acute High Blood Sugar And Blood Clotting

Sampson MJ, et al. "Monocyte and Neutrophil Adhesion Molecule Expression During Acute Hyperglycemia and After Antioxidant Treatment in Type 2 Diabetes and Control Patients." Arteriosclerosis, Thrombosis, and Vascular Biology. Volume 22, Pages 1187-1193. July 2002. http://atvb.ahajournals.org/cgi/content/full/22/7/1187

On the cellular membrane, there are receptors which contact and stick onto the cellular membrane of other cells. With the help from the cellular membrane adhesion molecules of white blood cells, neutrophils and monocytes contact the cellular membrane of the endothelial (the inner layer of an artery) and begin the process of atherosclerosis. *Atherosclerosis* is a disease in arterial blood vessels of various sizes. It is a result of chronic inflammation of the artery in which the inner wall is deposited with cholesterol plaques; sometimes, calcification on the outer wall of the artery occurs. *Arteriosclerosis* is the term used for atherosclerosis in a medium or larger artery. Arterio*lo*sclerosis is the terms used for atherosclerosis in a smaller artery.

Chronic inflammation can cause atherosclerosis; acute inflammation can cause blood clot formation. Blood clot(s) can block the blood vessel on the site or in other location(s) of the blood vessel(s) and block the circulation. This serious situation is the cause of heart attack, stroke, gangrene of the limbs and bowels, and necrosis of other internal organs. Both monocytes and neutrophils have several factors that promote their cells sticking onto the wall of the arteries. The factors, on which the following study focused, were Mac-1and others.

This study investigated the link between an acute increase of blood glucose level and the increase of inflammatory factors in the blood circulation. The study used a bolus of 75 grams of glucose by mouth to boost the blood glucose level, like that for glucose tolerance test. The results showed that the bolus of glucose caused a significantly high increase of monocyte Mac-1 in both the diabetic patients and the non-diabetics. Mac-1 is highly inflammatory and prothrombotic (promoting blood clot formation). The study showed a strong link between an acute load of blood glucose and an increased risk of vascular episodes or accidents, even in a non-diabetic. To avoid eating lots of carbohydrates at any time is the best way to prevent heart attack, stroke, and acute problems as a result of blood clotting that blocks the circulation.

Nitric Oxide And Vasodilatation

Scherrer U, et al. "Nitric oxide release accounts for insulin's vascular effects in humans." Journal of Clinical Investigation, Volume 94, Number 6, Pages 2511–2515. December 1994. http://www.pubmedcentral.nih.gov/articlerender.fcgi?tool=pmcentrez&artid=330085

The endothelium (the inner layer of the blood vessels) produces nitric oxide (NO), which relaxes the artery in a process called vasodilatation. Consequently, nitric oxide can increase the blood flow and lower the blood pressure. We already know the role of insulin in the metabolism (use and disposal) of blood glucose. We also found that insulin has a role in vasodilatation. This study concluded that insulin helped stimulate the release of nitric oxide by the endothelium. Abnormality of this mechanism was responsible for vasoconstriction and hypertension in the condition of (diabetic) insulin resistance.

High Blood Glucose And Nitric Oxide

Giugliano, D, et al. "Vascular Effects of Acute Hyperglycemia in Humans Are Reversed by L-Arginine. Evidence for Reduced Availability of Nitric Oxide During Hyperglycemia." Circulation, of the American Heart Association, Volume 95, Number 7, Pages 1783-1790. 1997. http://circ.ahajournals.org/cgi/content/full/95/7/1783

This study investigated how high blood glucose could

cause shrinking of the blood vessels (especially the arteries) and a rise in the blood pressure. This study reported that an acute high concentration of blood glucose (270 mg%) resulted in significant increases in both systolic and diastolic blood pressures, heart rate, blood viscosity (the blood is getting thicker and stickier, and moving slower), and circulating catecholamines (which are hormones for increasing blood pressure). The study also observed decreases in the leg blood flow and aggregation of platelets (which can help blood clot). The study suggested that high blood glucose reduced the availability of nitric oxide, increased vascular tone, blood pressure, blood viscosity (stickiness or thickness), and decreased blood flow.

High Blood Sugar Initiates Atherosclerosis

Marfella R, et al. "Circulating Adhesion Molecules in Humans. Role of Hyperglycemia and Hyperinsulinemia." Circulation, of the American Heart Association, Volume 101, Number 19, Pages 2247-2251. 2000. http://circ.ahajournals. org/cgi/content/full/101/19/2247

- Bonding of the white blood cells (monocytes and neutrophils) to the inner wall of an artery starts the process of atherosclerosis. Atherosclerosis is a result of deposits of LDL-cholesterol, especially VLDL-cholesterol, on the vascular wall. It narrows the lumen of the vessel, hardens the vascular wall, damages and weakens the structure of the vascular wall, and creates aneurysms. Atherosclerosis may also result in blood clots inside a vessel or the rupture of the aneurysm. There are factors that promote the adhesion of the white blood cells to the vascular wall, including the presence of a soluble intercellular adhesion molecule-1 and vascular adhesion molecule-1. (Please see ACUTE HIGH BLOOD SUGAR AND BLOOD CLOTTING.) This study showed a positive link between an acute increase in the concentration of circulating blood glucose (294 mg%) and the increase in the concentration of the soluble intercellular adhesion molecule-1 in both the healthy and the complication-free type 2 diabetics. However, the changes were more pronounced in the cases of type 2 diabetics. The study also found a positive link between reduction in the concentration of the circulating insulin and the increase of the soluble intercellular adhesion molecule-1, especially in cases of the complication-free type 2 diabetics.

Acute Hyperglycemia And Oxidation

Marfella, UR, et al. " Acute hyperglycemia induces an oxidative stress in healthy subjects." Journal of Clinical Investigation, Volume 108, Number 4, Pages 635–636. August 15, 2001. http://www.pubmedcentral.nih.gov/articlerender.fcgi?tool=pmcentrez&artid=209408

As mentioned, the endothelium (inner layer of the blood vessels) produces and releases nitric oxide (NO), which is the main factor for relaxing the blood vessels or causing vasodilatation (opening of the blood vessel). Consequently, the release of nitric oxide increases the blood flow and lowers the blood pressure.

Studies have shown that continuing to control the blood glucose level in diabetic patients improves their vascular complications, in both micro (local) and macro (total) stages. Microvascular changes by increased blood glucose may well activate the endothelium from a state of quiet, relaxed, anti-thrombosis (against the formation of blood clot), anti-adhesion (against sticking to the endothelium by cells such as monocytes), anti-oxidation (against an increase in the oxidative number of a molecule, atom or ion), to a state of generating reactive oxidative species (a molecule, atom or ion which promotes oxidation), of constricting the blood vessels, of pro-thrombosis (promoting blood clot formation), of pro-adhesion (promoting cells, such as monocytes, to stick to the endothelium) and other atherosclerotic risks.

Pennathur and coworkers reported; that hyperglycemia promotes oxidative reactions in the artery wall in monkeys. "A hydroxyl radical-like species oxidizes cynomolgus monkey artery wall proteins in early diabetic vascular disease. ("The Journal of Clinical Investigation. Volume 107, Number 7, Pages 853-60. April 2001. http://www. pubmedcentral.nih.gov/articlerender.fcgi?tool=pubmed&pub medid=11285304)

Nitrotyrosine is a molecule increased in the blood circulation, when the blood vessel is in a process of oxidation. This study investigated the relationship between the level of plasma nitrotyrosine; and the concentration of blood glucose. The study observed that the higher the blood glucose, the higher the concentration of nitrotyrosine and the tendency of oxidation of the blood vessel would be. The study also pointed out that studies have shown acute high blood glucose (hyperglycemia) can cause acute thrombosis and vasoconstriction, with abnormal changes in electrocardiogram (EKG). Also, other studies have also found microscopic (tiny, detailed or minuscule) damages to both the heart and vessels by high blood glucose level.

Acute High Blood Sugar And Inflammation

Esposito K, et al. " Inflammatory Cytokine Concentrations Are Acutely Increased by Hyperglycemia in Humans Role of Oxidative Stress." Circulation, Published online Volume 106, Number 16, Pages 2067-2072. September 30, 2002, http://circ.ahajournals.org/cgi/content/full/106/16/2067

During recent years, more studies have shown the role of inflammation in the process of developing many more diseases. While we are continuing to learn about the factors, which take part in the process of inflammation that eventually develops diseases, we also want to learn more about other factor(s) that may trigger the production and release of the causes of inflammation.

Hyperglycemia is one of the most important factors that promote inflammation. The levels of interleukin-6 (IL-6) and tumor necrosis factor--α (TNF--α) are increased in the blood circulation of the cases of diabetes mellitus. Also, interleukin-18 (IL-18) is increased in acute hyperglycemia, following acute myocardial infarction. IGT (impaired glucose tolerance) is a result of abnormal glucose level in a glucose challenge test, because insulin from the pancreas cannot bring the glucose level to normal.

This study found that, in people either with or without impaired glucose tolerance test (IGT), an acute hyperglycemia resulted in a significant increase of the IL-6 and TNF--α levels. The increase of the IL-6 and TNF--α levels lasted longer with those with IGT. The study also observed that the use of glutathione (an antioxidant) was effective in suppressing the increases of IL-6 and TNF--α. This result suggested that high blood glucose might play a role in causing diabetes through inflammation.

Egg, Carbohydrate Restriction, And Inflammation

Ratliff JC, et al. "Eggs modulate the inflammatory response to carbohydrate restricted diets in overweight men." Nutrition & Metabolism. Volume 5, Number 6. March 2008. http://www.nutritionandmetabolism.com/content/5/1/6

One thing that we did not know about cholesterol, was that most of the cholesterol inside the body is produced by the body from glucose. (Please see EGGS AND THE RISKS OF CARDIOVASCULAR DISEASE.) The researchers of this study previously found that men who ate carbohydrate-restricted diets had improved their dyslipidemia (abnormally high level of lipids or fats in the blood circulation) and had lowered blood glucose and insulin. Men on a carbohydrate-restricted diet had increased HDL-c, only if they ate cholesterol. This present study looked into the effects of egg on other biomarkers for inflammation. Inflammation is the focus of many diseases of our body. The study compared the impacts on inflammation by two identical carbohydrate-restricted diets containing cholesterol, except one diet included eggs. Both diets resulted in loss of weight and total body fat as well as trunk (waistline) fat mass. The diet including eggs caused a significant decrease in plasma C-reactive protein (CRP). CRP is a protein, produced by the liver in response to inflammation, and reflects the level of inflammation.

The study concluded that eating eggs could decrease plasma CRP, probably because of the presence of more cholesterol. Decrease of plasma CRP help control the response to inflammation.

CARBOHYDRATES AND CARDIOVASCULAR DISEASES

Fasting Blood Sugar And Risk Of Cardiovascular Diseases

Asia Pacific Cohort Studies Collaboration. "Blood Glucose and Risk of Cardiovascular Disease in the Asia Pacific Region." Diabetes Care Volume 27, Number 12, Pages 2836-2842. 2004. http://care.diabetesjournals.org/cgi/content/full/27/12/2836

This studies' collaboration found that the lower the value of fasting blood glucose, the lower the risk of cardiovascular disease would be. The positive link continued, until the level of fasting blood glucose dropping to at least 94.08 mg% and below. (For your information: the normal fasting blood glucose is between 70-100 mg%.) The study also showed that, in both men and women, reducing every scale of 19.2 mg% in fasting blood glucose would reduce by 21% of the risk of stroke and by 23% of the risk of ischemic heart disease.

Blood Glucose Levels And Risks Of Cardiovascular Diseases

Meigs JB, et al. "Fasting and Postchallenge Glycemia and Cardiovascular Disease Risk: The Framingham Offspring Study." Diabetes Care, a journal of the American Diabetes Association, Volume 25, Number 10, Pages 1845-1850, 2002. http://care.diabetesjournals.org/cgi/content/abstract/25/10/1845

The levels of fasting blood sugar (glucose) and the two-hour post-challenge blood sugar are the criteria for diagnosis with diabetes (DM). The fasting blood sugar is the value of the individual's blood sugar, after fasting for 8 to 10 hours. The two-hour post-challenge blood sugar is the value of the individual's blood sugar at the end of a two-hour period, after the individual has taken 75 grams of glucose by mouth, and after an initial fasting for 8 to 10 hours before the challenge. The criteria for diabetes (DM), according to the 1997 American Diabetes Association, is a level of fasting blood sugar at or more than 126 mg% (126 mg per 100 ml of blood). The criteria set by the World Health Organization include the levels of both the fasting blood sugar higher than 126 mg%; and the post-challenge blood sugar higher than 200 mg%.

This study investigated the relationship between the values of the fasting blood glucose and two-hour post-challenge blood glucose and the risk of cardiovascular disease. It found that an abnormal two-hour-post-challenge blood glucose level (higher than 200 mg%) was an independent risk factor for cardiovascular diseases. The risk is 1.1 times higher for an increase of every 38.18 mg% (2.1 mmol/L) of the two–hour post-challenge, over 200 mg/%.

High Blood Sugar Causes Vascular Damages

Stratton IM, et al. "Association of glycaemia with macrovascular and microvascular complications of type 2 diabetes (UKPDS 35): prospective observational study." British Medical Journal, Volume 321, Number 7258, Pages 405–412. August 12, 2000. http://www.bmj.com/cgi/content/abstract/321/7258/405

Type 2 diabetic patients frequently develop cardiovascular complications. It is important to find out the level of blood sugar for an extended period of time and the risk of vascular damages, including both the macroscopic (total) and microscopic (local) complications.

This study analyzed the data for the relationship between the level of hemoglobin A_{1c}, and the vascular (blood vessel) complications such as diabetes-related death, myocardial infarction (death of the heart muscles), stroke, amputation (cutting off a part or total of a limb) as a result of peripheral vascular damages, retinal (eye-ground where starts our eyesight) arterial damages, cataract formation and extraction (removal), and heart failure.

The study confirmed a significant relationship between the high level of blood glucose and the vascular complications. It also found that reducing 1% of Hemoglobin A_{1c} (35mg% of blood glucose) would reduce 21% of the risk of reaching the end-stage of diabetes and the diabetes-related death, 14% for myocardial infraction, and 37% for small vascular damages, such as retinal bleeding (bleeding from the eye ground). The study suggested that those with a level of Hemoglobin A_{1c} below 6.0% (135 mg% of blood glucose or below) had the lowest risk of vascular damages.

Type I Diabetes With Damages To Blood Vessels

Romney JS and Lewanczuk RZ. "Vascular Compliance Is Reduced in the Early Stages of Type 1 Diabetes." Diabetes Care, Volume 24, Issue 12, Pages 2102-2105. December 2001. http://care.diabetesjournals.org/cgi/content/abstract/24/12/2102

The heart contracts to push blood into the aorta and the peripheral arteries. After each contraction, the heart relaxes and eases the blood flow. The blood vessel is expanded when the blood flows forward, and it contracts when the blood flow eases. Therefore, the blood flow moves in waves; so does the pulse when we check it. Compliance of a blood vessel is the elasticity of its wall in expansion and contraction. Measurement of the pulse-wave of a blood vessel shows the compliance of the blood vessel. A blood vessel loses compliance or suffers microvascular complications when its wall is hardened and/or damaged.

This study compared the compliance of large and small blood vessels, in both patients with type 1 diabetes (DM) and the healthy non-diabetic. The study suggested that type 1 diabetic (DM) patients had vascular changes (loss of vascular compliance) early on, before they were clinically found microvascular complications. *Should not we suspect that vascular changes have started early on, as a result of chronic inflammation from the repeated high blood glucose after each meal? By the time we are diagnosed with diabetes (DM), our blood vessels have already been damaged.*

High Blood Sugar Damages The Small Blood Vessels

Manaviat MR, Afkhami M and Shoja, MR. "Retinopathy and microalbuminuria in type II diabetic patients." Ophthalmology of BMC, Volume 4, Number 9. 2004. http://www.pubmedcentral.nih.gov/articlerender.fcgi?tool=pmcentrez&artid=459228

Retinopathy is a pathological state of the eye ground (retina), which commonly shows damages to the small arteries, with bleeding and loss of vision. Retinopathy often happens in diabetic (DM) patients. *Microalbuminuria* is a positive finding of albumin (one of the blood or plasma proteins) in the urine test. Microalbuminuria often appears in patients who have suffered damages to the small arteries of their kidneys by diseases, such as diabetes (DM) and hypertension (high blood pressure), among others.

This study found that the degree of retinopathy was worse in patients with a lower body mass index. The patient with severe retinopathy had higher hemoglobin A_{1c} than those who did not have retinopathy. The study also found microalbuminuria was a significant, positive indicator, for diabetic retinopathy in case of type 2 diabetes (DM).

The patient with advanced stage of diabetes (DM) usually loses weight, because he does not have enough insulin to convert glucose into fat for storage. A person whose body mass index is either within or below the normal range has a greater possibility of becoming diabetic (DM) without a slightest knowledge of the undergoing disease process. He probably takes in the amount of calories enough for running his daily activities. However, he takes in mostly carbohydrates for his daily supply of calories.

Low Glycemic Index Carbohydrates And Coronary Heart Disease

Brand-Miller JC. "Glycemic index in relation to coronary disease." Asian Pacific Journal of Clinical Nutrition, Volume 13, Supplement S3. 2004.

Professor J. Brand-Miller pointed out that we had paid all the attention to dietary fat in the studies of cardiovascular disease and neglected the undesirable role of carbohydrates in our diet. Studies showed greater risks for the cardiovascular disease and its related mortality by the magnitude of postprandial hyperglycemia (high blood sugar after meal), even among the non-diabetic population. Large studies also found patients with high blood sugar in a glucose tolerance test was linked to a 1.8 to 3 times higher risks of death. The findings underscored that controlling the glycemic effects of dietary carbohydrates was important in preventing and managing the risks of coronary heart disease.

Brand-Miller also pointed out that diets with high-glycemic-index carbohydrates caused an increase in the daylong levels of both blood sugar and insulin, worsened insulin resistance in susceptible patients, negatively impacted the indicators for monitoring and managing metabolic syndrome, and threatened the health of individuals with a greater odd of coronary heart disease. Brand-Miller cited that higher blood glucose, even within the normal range, could still work against the endothelial function through different procedures, such as increasing oxidation (reducing the production and release of nitric oxide), increasing the inflammatory factors, promoting protein glycation, increasing LDL oxidation, growing pro-coagulatory (initiating blood coagulation and building blood clots), and rising anti-fibrinolytic activity (strengthening or hardening the blood clot).

In other studies, a positive link was found between a diet with low glycemic index and the decreases in blood triacylglycerides and LDL-cholesterol in men. Also a positive association was found between the use of a low glycemic diet for merely four weeks and the improvement in insulin sensitivity and glucose uptake in the fatty tissue in women.

The Nurse Health Study, a large-scale research project, showed those who ranked the highest in consuming diets with high glycemic indices and glycemic load were at almost twice the risk of myocardial infarction (death of the heart muscles) as those who ranked the lowest in doing so. (Please see HIGH GLYCEMIC LOAD AND COROANRY HEART DISEASE.) Several studies found a consistent tie between diets with high glycemic index and lower levels of HDL cholesterol. In postmenopausal women, high glycemic index diets were closely tied to high levels of C-reactive protein (a marker of chronic low grade of inflammation) and higher risks of coronary heart disease.

High Blood Sugar Damages Blood Vessels

Selvin E, et al" Glycemic Control and Coronary Heart Disease Risk in Persons With and Without Diabetes." Archives of Internal Medicine, Volume 165, Number 16, Pages1910-1916. September 12, 2005. http://archinte.ama-assn.org/cgi/content/abstract/165/16/1910?etoc

After continuing to blame too much cholesterol and fat in the body as the main reasons for developing coronary heart disease for many decades, more recent researches have found too much blood sugar in our body is the cause for this serious health problem. The reading of Hemoglobin A_{1c} is an important indicator for the level of the blood sugar of our body. The percentage point of Hemoglobin A_{1c} goes up as the level of blood sugar rises.

Although in layman's terms "heart attack" includes many types of heart diseases, we have dealt most commonly with the problems of the arteries running inside the heart muscles. These arteries are the coronary arteries. The build-up of cholesterol on their walls and/or blood clotting can reduce or block the blood flow through the coronary arteries. When the arteries become narrowed or blocked, the heart muscles cannot get enough blood flow for supplying oxygen and nutrition and removing carbon dioxide and waste. The heart muscles will begin to ache and then die.

This study found that the higher the value of Hemoglobin A$_{1c}$ the higher the risk of coronary heart disease would be in both the diabetics and non-diabetics. The study also found that there was no risk for coronary heart disease; for those of the non-diabetics (no DM), when their level of hemoglobin A$_{1c}$ fell at 4.0% or below. *The blood glucose level is at 65 mg% or 3.5 mmol/, when hemoglobin A$_{1c}$ is 4.0%. The normal fasting blood glucose is between 70-100 mg%, when hemoglobin A$_{1c}$ is 5.0% or below. This study implied that one is not necessarily free of risk for coronary heart disease even if he has not been diagnosed with diabetes (DM), based on today's criteria for diabetes (DM). The way to reduce the risk for coronary heart disease is to control the level of blood sugar (glucose) by consuming fewer carbohydrates, especially with those foods with high glycemic indices.*

High Blood Sugar In Heart Attacks

Das UN. "Is insulin an endogenous cardioprotector" Critical Care, Volume 6, Number 5, Pages 389–393. July 31, 2002. http://www.pubmedcentral.nih.gov/articlerender.fcgi?artid=137317

When we are under stress, our body responds by triggering a release of hormones from organs to raise our blood glucose level. These organs and their hormones are the thyroid gland and thyroid hormone, the cortex of the adrenal glands and corticosteroid hormones, the medulla of the adrenal glands and adrenaline, as well as the pancreas and glucagon.

While hyperglycemia or high blood sugar under stress helps the body to fight or fly, it can cause several unwanted consequences. One of the unwanted consequences is the initiation of acute vascular changes as a result of inflammation and suppression of the production and release of nitric oxide from the endothelium (the inner wall of the blood vessel).

This report cited a positive relationship between stress hyperglycemia and diabetes mellitus with myocardial infarction, and the risk of death from congestive heart failure or shock as a result of a failing heart. The report also cited the consequences of hyperglycemia that caused inflammation and promoted endothelial dysfunction after triggering a cascade of factors. The report suggested that to improve the rate of recovery, without complication from acute myocardial infarction and critical illness, it was most important to maintain a blood sugar level at or below 110 mg%. A stable blood sugar level could be achieved by using insulin properly. *But, how about by dieting?*

Hyperglycemia, CRP, And Prognosis For Heart Attack

Wong, VW, et al. "C-Reactive Protein Levels Following Acute Myocardial Infarction: Effect of insulin infusion and tight glycemic control." Diabetic Care, Volume 27, Number 12, Pages 2971-2973, 2004. http://care.diabetesjournals.org/cgi/content/full/27/12/2971

Hyperglycemia (high blood glucose or sugar) triggers a cascade of inflammatory factors, including Interleukin-6 (IL-6), tumor necrosis factor-α (TNF-α), interleukin-18 (IL-18),

C-reactive protein (CRP), and others. After acute myocardial infarction, the death and recovery rates are closely related to how insulin therapy is provided to lower the blood glucose level. This study found that maintaining a normal blood glucose level with intravenous insulin-based therapy for the first 24 hours after acute myocardial infarction would slow down the increase of the CRP level. The study suggested that the lowered blood glucose level was responsible for the suppression of inflammatory response to myocardial infarction. *Right on target! I agree.*

CRP And Prognosis For Heart Attack

Bursi F, et al. " C-Reactive Protein and Heart Failure After Myocardial Infarction in the Community." The American Journal of Medicine, Volume 120, Number 7; Pages 616-622. July 2007. http://www.sciencedirect.com/science?_ob=ArticleURL&_udi=B6TDC-4NKB1YW-2&_user=10&_rdoc=1&_fmt=&_orig=search&_sort=d&view=c&_acct=C000050221&_version=1&_urlVersion=0&_userid=10&md5=dc96e15b7451b7b9d8162af1aa59aab6

When one falls seriously ill, everyone wants to know the odds for his recovery (prognosis). It would be very wonderful to have something for measuring the prognosis. This study found a strong positive link between the level of CRP (C-reactive protein) and the odds for developing heart failure or death after suffering acute myocardial infarction. Therefore, the higher the CRP value was at the time when myocardial infarction started, the higher the rate for developing heart failure or facing death during the course of the disease. CRP is a protein from the liver and reflects the level of inflammation inside the body. As reported in other studies, hyperglycemia promotes the increase of CRP. Tight control of blood sugar after myocardial infarction should improve the prognosis.

High Blood Sugar And Atherosclerosis

Aronson D and Rayfield, EJ. "How hyperglycemia promotes atherosclerosis: molecular mechanisms." Cardiovascular Diabetology, Volume 1, Number 1. 2002. http://www.cardiab.com/content/1/1/1

This study found that blood glucose (fructose and galactose too) could

- simply bond with protein or fat to become glycoprotein or glycolipid (the bonding process is called glycation) which, in turn, disrupts the normal molecular functions and promotes atherosclerosis;
- promote oxidation (that can interfere the synthesis and release of nitric oxide);
- activate the production and release of C-reactive protein from the liver (that reflects the state of inflammation). The three mechanisms could promote each other.

CARBOHYDRATES AND RESPIRATORY DISEASES

Actors and actresses smoking in the movies made the habit terribly fashionable up until recent years. At the same time we saw a dramatic increase in respiratory diseases, including inflammation, allergies, and cancers. We have blamed many things as the causes for our respiratory diseases. But only a very few of us have realized that carbohydrate diets play an important role in making problems for our respiratory system.

Excessive Carbohydrate Load Causing Breathing Failure

Covelli HD, Black JW, Olsen MS and Beekman JF. "Respiratory failure precipitated by high carbohydrate loads." Annals of Internal Medicine, Volume 95, Number 5, Pages 579-81. November 1981. http://www.ncbi.nlm. nih.gov/sites/entrez?cmd=Retrieve&db=PubMed&list_uids=6794409&dopt=Citation

In the body, glucose is used primarily as a source of energy. Carbohydrates are the source of glucose. In energy production, our body uses up a unit of oxygen molecule to burn carbohydrates and produce a unit of carbon dioxide (molecule). On the other hand, our body uses up one unit of oxygen molecule to burn fat and produces only 0.7 unit of a carbon dioxide molecule, and consumes a unit of oxygen molecule to burn protein and produces 0.8-0.9 unit of carbon dioxide molecule.

The body has to constantly release carbon dioxide, mainly through our respiratory system, and some through the skin. If the body fails to release carbon dioxide effectively for any reason, we suffer respiratory failure and need emergency care. This may well include a breathing machine to help us breathe and get the carbon dioxide out.

This study reported that three patients urgently needed breathing machines to help them breathe, after taking high amounts of carbohydrate intravenously for total nutritional support only a few hours earlier. Those patients produced too much carbon dioxide, after their bodies used the overloaded carbohydrates. Then, the patients' arterial blood accumulated carbon dioxide. The study concluded that too much carbohydrate loading might cause respiratory acidosis in patients who were unable to adequately improve their alveolar ventilation to compensate for increased carbon dioxide production. *Respiratory acidosis* is the blood to become more acidic, as a result of failing to remove carbon dioxide through breathing. *Alveoli are* the terminal space of the respiratory tract where the circulating blood takes in oxygen and releases carbon dioxide.

Low Carbohydrate Diets Improve COPD

Angelillo VA, et al. "Effects of low and high carbohydrate feedings in ambulatory patients with chronic obstructive pulmonary disease and chronic hypercapnia." <u>Annals of Internal Medicine</u>, Volume 103, Number 6 (Part 1)), Pages 883-5. December 1985. http://www.ncbi.nlm. nih.gov/sites/entrez?cmd=Retrieve&db=PubMed&list_ uids=3933397&dopt=Abstract

The body produces more carbon dioxide when it uses more carbohydrate to produce energy for its activities. This study showed the low-carbohydrate diet that contained 28% of the total calories from carbohydrates, 55% from fats, helped the patients with chronic obstructive pulmonary disease (COPD), significantly reducing the production of carbon dioxide and their arterial blood level of carbon dioxide. That improved respiratory acidosis. The study also reported a significant improvement in the patients' lung function tests. The daily calorie intake, with the low carbohydrate diet, was more than the patient's daily calorie consumption. This study suggested, a low-carbohydrate, high-fat mixture might be beneficial for patients with COPD.

Low Carbohydrate Diets Improve Respiratory Failure

Kwan R and Mir MA. "Beneficial effects of dietary carbohydrate restriction in chronic cor pulmonale." American journal of Medicine, Volume 82, Number 4, Pages 751-8. April 1987. http://www.ncbi.nlm.nih.gov/sites/entrez?cmd= Retrieve&db=PubMed&list_uids=3105310&dopt=Citation

Both consuming too much oxygen and producing too much carbon dioxide can cause both respiratory failure and heart failure. Reducing the consumption of carbohydrates produces less carbon dioxide and consumes less oxygen by the body. In this study, its researchers cut the amount of carbohydrates of the diets, but kept the same amount of calories of the participants' original diet before the study. They used two carbohydrate-reduced diets that contained 200 grams and 50 grams of carbohydrates respectively. They enrolled eight participants in the study.

This study suggested that in short-term, hospital-controlled situations, reducing the patients' carbohydrate intake to 200 grams a day improved the general well-being of those patients who had chronic hypercapneic (too much carbon dioxide in the arterial blood) and respiratory failure by increasing arterial oxygen tension (more oxygen in the blood) and decreasing arterial carbon dioxide tension (less carbon dioxide in the blood). Reducing the patients' carbohydrate intake to 50 grams a day provided even more beneficial effects. Such a diet might be used in patients who had intractable (stubborn) respiratory failure.

Carbohydrate And Acute And Chronic

Pulmonary Diseases

Malone AM. "The Use of Specialized Enteral Formulas in Pulmonary Disease." Nutrition in Clinical Practice, Volume 19, Number 6, Pages 557-562. 2004. http://ncp. aspenjournals.org/cgi/content/abstract/19/6/557

This review report pointed out the serious impacts on the outcomes in the patient's pulmonary disease resulting from malnutrition and its ill consequence in pulmonary functions. Dr. Malone said that the understanding of worsening respiratory functions as a result of high carbohydrate in the infusion nutritional therapy (parenteral feeding or intravenous feeding) had created a controversy in using an alternative high-fat nutritional supply through the gastrointestinal tract (enteral feeding).

The review report suggested that the nutritional supply, with less carbohydrate to the patient who had pulmonary disease, resulted in a lower production of carbon dioxide. In the case of acute respiratory distress syndrome, an inflammatory state, an enteral formula with modified lipids would favorably influence the inflammatory mechanism. The report suggested, "Severely malnourished ambulatory COPD patients who need repletion (satiety) are more likely to benefit from a high-fat formula than a similar normally nourished patient."

CARBOHYDRATES AND DIGESTIVE DISORDERS

Hyperglycemia And Oral Inflammation

Ship JA, "Diabetes and oral health." Journal of American Dental Association, Volume 134, Number, Pages 4S-10S.

Supplement 1, 2003. http://jada.ada.org/cgi/content/full/134/suppl_1/4S

The mouth is always in contact with bacteria, viruses, and various other unclean materials because it is open to outside the body. At one time or another, we have experienced pain inside our mouth, including the gums, teeth, tongues, tonsils, etc., as a result of infection, irritation, or both. Eventually this may have led to inflammation. We have known that people with diabetes (DM) have a high risk of oral infection. We must also realize that people with higher blood glucose levels, and without a diagnosis of diabetes (DM), also suffer frequent inflammation of the mouth.

This article pointed out that diagnosis with diabetes (DM) was often made based on numerous systemic and oral signs and symptoms, including gingivitis (inflammation of the gum), periodontitis (inflammation around the tooth), recurrent oral fungal infections, and impaired wound healing. This article added that diabetes (DM) caused alveolar bone loss (alveolar bone is the jaw bone where holds tooth socket), periodontitis-induced bacteremia (bacterial infection inside the blood circulation), dental caries, salivary dysfunction (dry mouth or xerostomia), bad taste, pain or burning sensation of the mouth or tongue, and oral mucosal diseases such as inflammation of the mouth.

Although this review article was focused on diabetic patients' dental and oral care, we must bear in mind that high blood glucose level happens way before diagnosis with diabetes mellitus is made. Oral/dental diseases can be very useful indications, for early detection and prevention of diabetes mellitus.

Sucrose And Bowel Inflammation

Persson PG, Ahibom A and Hellers G. "Diet and inflammatory bowel disease: a case-control study." <u>Epidemiology</u>. Volume 3, Number 1, Pages 47-52. January 1992. http://www.ncbi.nlm.nih.gov/sites/entrez?cmd=Retrie ve&db=PubMed&dopt=AbstractPlus&list_uids=1313310

Irritable bowel syndrome and Crohn's disease are well-known gastrointestinal diseases. They can be very difficult cases to completely cure. It is very important to find out the cause(s) of the diseases. One thing we know is that the bowel becomes inflamed. This study found that the risk of Crohn's disease increases by 2.6 times for the patients who consumed more than 55 grams of sucrose every day. The risk of Crohns' disease dropped to less than 50% for individuals who consumed more than 15 grams of fiber every day. The study also found that individuals who ate fast food more than twice weekly had a risk of Crohn's disease to 3.4 times higher than the individuals who did not. In the same situation, the risk of ulcerative colitis (inflammation of the large bowel with ulcers) among these people went up by 3.9 times more than those who did not eat fast food.

There were reports on using low starchy foods for treating and controlling Crohns' disease and ankylosing spondylitis.

Carbohydrates And Gallstones

Tsai G-J, Leitzmann MF, Willett WC and Giovannucci EL. "Dietary carbohydrates and glycaemic load and the incidence of symptomatic gall stone disease in men." <u>Gut</u>. Volume 54, Number 6, Pages 823-828. 2005. http://gut.bmj.com/cgi/content/abstract/54/6/823

For the past decades, we have blamed fats for causing gall bladder diseases, including the formation of gallstones. Is this really accurate? This study found that the higher the amount of daily carbohydrate consumption, the higher the risk of forming gallstones would be. It also observed that the higher the individual's glycemic load (the product of the total amount of food and its glycemic index), the higher the risk of gall bladder problems would be.

The study also found a positive relationship between the risk of forming gallstones and the amount of daily consumption of sucrose, fructose, and starch. They pointed out that men who took lots of carbohydrates with a high glycemic load and other foods with a high glycemic index had a high risk of forming gallstone disease. The article suggested that the results of their study indicated low-fat high-carbohydrate diets might not be "an optimal dietary recommendation."

CARBOHYDRATES AND ARTHRITIS

Carbohydrate Restriction And Gouty Arthritis

Dessein PH, et al. "Beneficial effects of weight loss associated with moderate calorie/carbohydrate restriction, and increased proportional intake of protein and unsaturated fat on serum urate and lipoprotein levels in gout: a pilot study." Annals of Rheumatic Disease, Volume 59, Pages 539-543, July 2000. http://ard.bmj.com/cgi/content/abstract /59/7/539

Gout is inflammation in the joints resulting from an increase of uric acid in the serum (the fluid of plasma without blood clotting factors). It often happens in the big toes. Doctors almost always advise patients to avoid certain foods, which can increase the amount of uric acid in the body and blood circulation. The disease involves inflammation, so we should look into the relationship between foods and the risk of gouty arthritis.

This article studied the relationship between carbohydrates, calorie intake, and the risk of gout. Increasing evidence showed that insulin resistance and abnormal blood lipoprotein-cholesterols had a role in causing gouty arthritis. Also, insulin resistance affected the excretion of uric acid from the kidneys. The study used a calorie-restricted diet (1,600 kcal daily), with lower carbohydrate (40% of the daily calories or 160 grams), higher protein (30% of the daily calories or 120 grams), and fat (30% of the daily calories or 54 grams). Carbohydrates in the diet were complex ones or polysaccharides. Fats in the diet were polyunsaturated fats and monounsaturated fats. Observation during the period of investigation revealed that the diet caused weight loss, reduction in the monthly frequency of gouty attacks and a drop from an average of 2.1 times to 0.6 times, and a decline in the amount of serum (blood) uric acid. The study also found a decrease in the levels of serum cholesterol, LDL-c (low density lipoprotein-cholesterol), triglyceride, and the ratio between serum cholesterol and HDL-c (high density lipoprotein-cholesterol). The study suggested that the dietary recommendations for gout might need reevaluation.

Starch And Ankylosing Spondilitis

Ebringer A and Wilson RC. "The Use of a Low Starch Diet in the Treatment of Patients Suffering from Ankylosing Spondylitis." Clinical Rheumatology, 1996, 15. Supplement

1:62-66. http://www.ncbi.nlm.nih.gov/sites/entrez?cmd=Retr
ieve&db=PubMed&list_uids=8835506&dopt=Citation

Ankylosing spondylitis is a painful back problem. Ankylosing means a joint or joints become fused and rigid with little or no range of movement. Spondylitis means a joint or joints of the spine have become inflamed and eventually degenerate. Ankylosing spondylitis has been thought to be a type of autoimmune disease, and probably a genetic disorder. However, research as to its causes and treatment has helped those suffering from it to experience a surprising relief from their back pain. Such treatment possibly has also helped either Crohn's or similar diseases.

HLA-B27 is Human Leukocyte Antigen B27. Leukocyte is a white blood cell. Antigen is a foreign protein from the microorganism that can trigger our immune system to produce antibody; therefore, our immune system can defend our body from the microorganism through antigen-antibody reaction. Vaccination is a good example for creating antibody by injecting weakened microorganism or its protein to stimulate the body for producing antibody. Thus, the body would have antibody react to the microorganism's antigen for defense. (Please see CARBOHYDRATES AND INFLAMMATION and CARBOHYDRATES AND IMMUNE SYSTEM.)

This article reported that a no-starch diet could relieve the individual's symptoms of ankylosing spondylitis. Studies showed Klebsiella, the bacteria found in the ankylosing spondylitis patients' bowel, had two molecules similar to the molecules of HLA-B27. In addition, an increasing amount of Klebsiella was found in stool samples of the ankylosing spondylitis patients and the patients who had lesions at the ileo-cecal junction, between the small intestine and the cecum (the beginning part of the large intestine). The patient who was diagnosed with Crohn's disease also has a lesion at the ileo-cecal junction. Klebsiella bacteria in the bowel need dietary starch for its growth. Cutting the starchy foods (a low-starch diet) showed a decrease in IgA (immunoglobulin A, an antibody for mucosal immunity such as defending the mucosal lining of the bowels) in both the healthy individuals and the ankylosing spondylitis patients. The low-starch diet had also reduced the symptoms of ankylosing spondylitis. The decrease in IgA reflects a decrease in the antigen from Klebsiella.

Blood Sugar And Rheumatoid Arthritis

Newkirk MM, et al. "Advanced glycation end-product (AGE)-damaged IgG and IgM autoantibodies to IgG-AGE in patients with early synovitis." Arthritis Research Therapies, 2003; 5(2): R82–R90. http://www.pubmedcentral.nih.'ov/ articlerender.fcgi?tool=pmcentrez&artid=165032

Glycation is a bonding between simple sugar (fructose, galactose, or glucose) and protein or lipid without the help from enzyme in a chemical reaction. A chemical reaction usually requires a catalyst and heat, as well as oxygen or enzyme(s) and so on. The endproduct of the bonding is called the advanced glycation endproduct (AGE). AGE is recognized as a cause of inflammation in the body.

Immunoglobulin G (IgG) can be damaged by AGE as a result of high blood glucose and/or damage of the cells and tissues by reactive oxidative species such as peroxide, superoxide, or hydrogen peroxide. The latter is a situation that the pro-oxidants prevail over the anti-oxidants. *(Continue to take antioxidants without first avoiding the buildup of oxidant is ineffective.)* An antibody to the AGE-damaged IgG (an autoimmune response) has been observed in patients who have suffered rheumatoid arthritis for a long period of time.

This study found the patients with arthritis had an antibody to the damaged-IgG by AGE, especially those with rheumatoid arthritis and swelling of the joint(s). Most importantly, the damaged-IgG by AGE was found in the early stage of disease among all the patients in this study.

CARBOHYDRATES AND THE IMMUNE SYSTEM

The body has the ability to defend itself with the immune system. Its immune system includes the innate immune system and the adaptive immune system. The *innate immune system* is a natural immune system, with which we were born. The innate immune system defends the body from invasion by foreign organisms, such as bacteria, viruses, and others. On the other hand, the *adaptive immune system* produces antibodies against specific antigens (protein, foreign to our body, from bacteria, viruses, and foods), after learning the nature of the antigen from the innate immune response. The response is repetitive and lasting.

The adaptive immune system can develop antibodies against the antigens from outside of the body. It can also misrecognize some proteins in the body as foreign to the body. In this case, the adaptive immune system develops antibodies against some proteins of an organ in the body, and destroys the organ. This is called autoimmune disease. *Developing antibody against the AGE-damaged IgG*

---

mentioned in the last article of the previous section is an example.

Hyperglycemia And The Defense Capability

Turina M, et al. "Acute hyperglycemia and the innate immune system: clinical, cellular, and molecular aspects." <u>Critical Care Medicine</u>, Volume 33, Number 7, Pages 1624-33. July 2005. http://www.ncbi.nlm.nih.gov/sites/en trez?cmd=Retrieve&db=pubmed&dopt=AbstractPlus&list_uids=16003073&query_hl=11

Our normal fasting blood glucose level falls within a range between 70 mg% and 100 mg%. A fasting blood glucose level over 100 mg% is called hyperglycemia (high blood glucose). When our fasting blood glucose level falls in the range between 100 mg% and 125 mg%, we will be diagnosed as pre-diabetic. If our fasting blood glucose level is at 126 mg% or higher, we will be diagnosed with diabetes (DM). After eating carbohydrates at each meal, our blood glucose level goes up. If our blood glucose level goes up higher than 180 mg% within two hours after meal, we are pre-diabetic. We consider we are not diabetic (DM) if our blood glucose level returns to or under 140 mg%, two hours after starting the meal.

Hyperglycemia (high blood glucose or sugar) plays an important role in affecting the normal functions of our body. One of them, which hyperglycemia seriously impacts, is the innate immune system. Our innate immune system or natural immune system immediately defends our body from tissue damage or infections by other organisms, such as bacteria, viruses, and so on, as soon as infections are started.

The innate immune system provides an immediate but a

short-term defense. The adaptive immune system, through an antigen-antibody reaction, provides a specific but a long-term defense against certain causes of diseases. What does our body's natural defense system do if we have a higher level of blood glucose shortly after eating lots of carbohydrates in a meal? This review article found that such a brief episode of acute hyperglycemia could significantly affect our innate immune system, and terribly weaken the ability of our body in defending itself against infection.

Hyperglycemia And Acute Inflammation

Esposito K, et al. "Clinical Investigation and Reports: Inflammatory Cytokine Concentrations Are Acutely Increased by Hyperglycemia in Humans: Role of Oxidative Stress." Circulation, Volume 106, Pages 2067. 2002. http://circ.ahajournals.org/cgi/content/full/106/16/2067

A glucose challenge test checks the changes in the patient's blood glucose levels after he takes 75 grams of glucose solution by mouth. If the blood glucose level fails to reach 140 mg% (140 milligram of glucose in 100 millimeter or cc. of blood or serum) or lower at the end of a two-hour period after taking the glucose solution, he is diagnosed with "impaired glucose tolerance" or IGT. All diabetic (DM) patients and some people who have not been diagnosed with diabetes (DM) demonstrate IGT.

Diabetic (DM) patients have higher levels of blood glucose (sugar), even eating their "ordinary" diets. The abnormally high levels of blood glucose are much more drastic after they take glucose challenge test, either because the patient's pancreas does not produce and release enough insulin to offset the excessive amount of glucose during the challenge, or because his tissue cannot use the excessive amount of glucose, even his pancreas has released a more-than-enough amount of insulin into the blood circulation. The latter situation is called "insulin resistance." (*Is not it possible that the pancreas of the diabetic patient perhaps produces defective insulin molecules that cannot help the tissue use glucose effectively?*)

Cytokines are proteins, which communicate between cells. Interleukin-6 (IL-6) and tumor necrosis factors (TNF) are cytokines. IL-6, TNF, and many other cytokines are increased in the tissues and blood circulation, when the body has inflammation as a result of tissue damage and/or infection. As an established fact, the blood circulation of diabetic patients has an increase of IL-6, TNF, and other cytokines.

This study found that acute hyperglycemia (high blood glucose level) increased the levels of circulating cytokines (IL-6, IL-18, and TNF for this study). It observed that hyperglycemia caused oxidation and then increased the inflammatory factors. It also observed that individuals with diabetes (DM) or impaired glucose intolerance (IGT) were more strongly affected by acute hyperglycemia.

Hyperglycemia And The Outcome Of Injury

Laird AM, et al. "Relationship of Early Hyperglycemia to Mortality in Trauma Patients." Journal of Trauma, Injury, Infection and Critical Care, Volume 56, Number 5, Pages 1058-1062, May 2004. http://www.jtrauma.com/pt/re/ jtrauma/abstract.00005373-200405000-00019.htm;jsessioni d=HXhN0hnJG1Dx92xZmFXv9Zlc1q9npHf74JnJvPmSzqf 9PBLTn8bC!65375592!181195628!8091!-1

High blood glucose (hyperglycemia) is undesirable. When is it so high that it is harmful to our health and life? This is an important question for medical researchers. During the recent years, we have found that a tight control on the patient's blood glucose level leads to a better outcome over the course of his treatment.

This study looked into the relationship between the outcomes of the patients' clinical course and their blood glucose levels, during the first one or two days (early hyperglycemia) of the patients' stay in the ICU (intensive care unit). The study set three cutoff points at 110 mg%; 150 mg%; and 200 mg%, for defining hyperglycemia. (1 mg% = 1 mg/dl. In this case, 1 mg% is 1 milligram of glucose in every 100 milliliter or cc of blood or plasma.) The study observed the following:

1. At the time of admission to the intensive care unit, the trauma patient's blood glucose level was at or higher than 200 mg/dl. This would have posed a significantly higher threat of infection and death regardless of injury characteristics.

2. The trauma patient's outcome was improved when his blood glucose level was below 200 mg/dl, at the time of admission to the intensive care unit.

3. Tight glucose control is necessary to keep serum glucose lower than 200 mg/dl for critically ill trauma patients. However, it may not be necessary to aggressively keep the patient's blood glucose level lower than110 mg/dl, as reported by other studies.

CARBOHYDRATES AND OSTEOPOROSIS

Osteoporosis and osteopenia are a state of bone disorder

with a loss of the mineral calcium from the bone. Based on the degree of mineral loss identified with a special X-ray technique, osteopenia is milder than osteoporosis. However, it is very important to prevent osteopenia and osteoporosis from developing or getting worse after the disease has started. Osteoporosis and osteopenia are most often found in the older population. These disorders may cause fractures of their back, limbs including joints, and other areas.

Up to this time, doctors have been using a calcium supplement and Vitamin D in attempt to improve the patient's bone density. However, this clinical approach may not be as effective as we have expected. We still need to know the real cause(s) of osteopenia and osteoporosis. After we have identified the cause(s) of the disorder(s), we may be able to prevent osteopenia and osteoporosis from happening in the first place.

Systematic Inflammation And Osteoporosis

Koh J-M, et al. "Higher circulating hsCRP levels are associated with lower bone mineral density in healthy pre- and postmenopausal women: evidence for a link between systemic inflammation and osteoporosis." Osteoporosis International. Volume 16, Number 10, Pages 1263-1271. October 2005. http://www.ncbi.nlm. nih.gov/sites/entrez?cmd=Retrieve&db=PubMed&list_ uids=15702263&dopt=AbstractPlus

Alkaline phosphatase (ALP), an enzyme in the body, usually becomes active and is increased in blood circulation when bone structure is losing its minerals including calcium. The level of the high sensitivity C-reactive protein (hsCRP) is closely related to the degree of inflammation inside the body.

This study investigated the relationships among the blood (or serum) level of alkaline phosphatase (ALP), the blood (or serum) level of high sensitivity C-reactive protein, and the bone mineral density to determine if inflammation was causing osteopenia and osteoporosis in the otherwise-healthy women. The study suggested that unrealized systemic inflammation might be related to the bone turnover rates and deterioration of the bone mass.

Inflammation, Aging, And Osteoporosis

Ginaldi L, Di Benedetto MC and De Martinis M. "Osteoporosis, inflammation and ageing." Immunity & Ageing, Volume 2, Number 14 2005. http://www. immunityageing.com/content/2/1/14

Osteoporosis has become an important medical disorder during recent decades. The patient with osteoporosis loses bone mass. His/her bones increasingly become fragile and easy to fracture. Older people have a higher risk of osteoporosis. They face a higher risk of fractures and disability. Keep it in mind that diabetic patients have higher rate of inflammation and osteoporosis.

More studies in recent years have suggested that activation of the immune system is involved during the course of osteoporosis. More importantly, some of these proinflammatory (promoting inflammation) factors were involved in regulating the activity of osteoblasts (responsible for bone formation) and osteoclasts (responsible for bone destruction). These proinflammatory factors promoted less bone formation, and more bone destruction, and caused osteoporosis. This report suggested that chronic inflammation and changes in the immune system while aging might be among the causes of osteoporosis.

Hyperglycemia And Fracture

Schwartz AV. "Diabetes Mellitus: Does it Affect Bone?" Calcified Tissue International, Volume 73, Number 6, Pages 515-9. December 2003. http://www.ncbi.nlm. nih.gov/sites/entrez?cmd=Retrieve&db=PubMed&list_ uids=14517715&dopt=AbstractPlus

Getting a fracture is quite common among members of the older population. Diabetes (DM) is commonly related in fractures of the hip, upper section of the humerus (upper arm bone), and foot. This study reported that both type 1 and type 2 diabetes (DM) patients have a high risk of fractures. Type 1 diabetes (DM) was found to have a fairly small reduction in bone mineral density (BMD). However, Type 2 diabetes (DM) was often found to have an increase in bone mineral density (BMD). It is a mystery as to why type 2 diabetes (DM) patients, with a higher bone mineral density (BMD), would still have a high risk of fractures. The report suggested that good understanding of the impact by diabetes (DM) on the bone structure would improve the outcome of preventing fractures in older diabetic patients.

Hyperglycemia And Bone Formation

Yasuda S and Wada S. "Bone metabolic markers and osteoporosis associated with diabetes mellitus." Clinical Calcium, Volume 11, Number 7, Pages 879-83. July 2001. http://www.ncbi.nlm.nih.gov/sites/entrez?cmd=Retrieve&db =pubmed&dopt=AbstractPlus&list_uids=15775593&itool=i conabstr&query_hl=5

Osteocalcin is a non-collagenous (not of the collagen

family) protein secreted by osteoblasts. Osteoblasts are responsible for bone formation. Osteocalcin is found in both bone and teeth. It is thought to play a role in balancing the levels of calcium and mineralization in the body. Osteocalcin could be bonded with glucose, thus lose its function of balancing the level of calcium and mineralization. The bonding is glycosylation or a process of glycation with the involvement of enzyme. A drop in the level of free osteocalcin in the circulation may be a result of more osteocalcin being glycosylated (glucose-bonded), in cases of poor glycemic control (control of blood glucose). For that, tight glycemic control increases the levels of osteocalcin thus decreasing the risk of fracture for the diabetic (DM) patients. An increase of blood level of osteocalcin is a biochemical marker for improvement of bone mineral density or bone formation in the course of treatment for osteoporosis.

This study reported an association between the metabolic effects of poor glycemic control and an increase in bone re-absorption (loss of calcium), while there was no bone formation at the same time. Reports showed the levels of osteocalcin decrease in cases of poor glycemic control among diabetic patients.

CARBOHYDRATES AND SEXUAL DYSFUNCTION
Hyperglycemia And Erectile Dysfunction

Basu A and Ryder REJ. "New Treatment Options for Erectile Dysfunction in Patients with Diabetes Mellitus." Drug, Volume 64, Number 23, Pages 2667-2688, 2004. http://www.ncbi.nlm.nih.gov/sites/entrez?db=pubmed&cmd=Retrieve&dopt=AbstractPlus&list_uids=15537369&query_

hl=4&itool=pubmed_docsum

Erection of the male penis and female clitoris requires not only a functioning nervous system but also expandable blood vessels. The latter is important, because the penile or clitoral erection is a result of dilation of the blood vessels of the penis or clitoris, which becomes engorged when blood pools in the vessels. Nitric oxide (NO) is the chemical substance for dilating the blood vessels. Nitric oxide is produced and released by the inner layer (endothelium) of the vessels. Nitric oxide activates an enzyme, guanylate cyclase, which produces cyclic guanosine monophosphate (cGMP) for engorgement of both the penis and clitoris. There is an enzyme, phosphodiesterase (PDES) type 5 in the penis (and clitoris?), which is responsible for breaking down eGMP. When there is a shortage of nitric oxide in the blood vessels of the penis, erection cannot happen, or it cannot sustain through the act of intercourse, thus it causes erectile dysfunction. Of course, there are other causes for erectile dysfunction.

This study cited that hyperglycemia (high blood glucose) interferes the production and release of nitric oxide (NO) from the endothelium. It disrupts the ordinary vascular responses in penile erection. Erectile dysfunction in cases of diabetes mellitus is strongly, positively linked to the degree of blood glucose control, length of the disease, and complications of diabetes mellitus. But, you don't have to become diabetic before you suffer erectile dysfunction!

Diabetes Mellitus And Erectile Dysfunction

Bacon CG, et al. "Association of Type and Duration of Diabetes With Erectile Dysfunction in a Large Cohort of

Men." DIABETES CARE, Volume 25, Number 8, Pages 1458-1463. August 2002. http://care.diabetesjournals.org/cgi/content/abstract/25/8/1458

Impotence or erectile dysfunction (ED) has been a problem with men who are 50 years and older. Because of personal pride, most men do not wish to discuss about the problem, even with their physician. In the meantime, physicians have realized that, besides contributing factors such as psychological and hormonal disorders, diabetes (DM) is the cause in the majority of cases of erectile dysfunction. This study found, in comparison to the non-diabetic men, type 1 diabetic men had a 3.0 times and type 2 diabetic men had a 1.3 times higher risk for ED. Among the type 2 diabetic men, those who had had a longer duration of diabetes (DM) had a higher risk for ED, than those who had had diabetes (DM for a shorter time.

Hyperglycemia, Erectile Dysfunction, And Cardiovascular Diseases

Grover SA, et al. "The Prevalence of Erectile Dysfunction in the Primary Care Setting: Importance of Risk Factors for Diabetes and Vascular Disease." Archives of Internal Medicine, Volume 166, Number 2, Pages 213-219. 2006. http://archinte.ama-assn.org/cgi/reprint/166/2/213

Erectile dysfunction shares the same cause(s) for many serious cardiovascular diseases (CVD) and it can often be used as a predictor for those diseases. For prevention or early detection of cardiovascular diseases, men should report changes in their sexual functions to their physician. This review article observed that in comparison to the patients who did not have erectile dysfunction, those who suffered erectile dysfunction had a 1.45 times higher risk of cardiovascular disease and a 3.13 times higher risk for diabetes (DM). The patients with ED would have higher risks for coronary heart disease, abnormal fasting blood glucose or undiagnosed hyperglycemia, and diabetes (DM).

Hyperglycemia, Erectile Dysfunction, And Coronary Heart Disease

Min JK, et al. "Prediction of Coronary Heart Disease by Erectile Dysfunction in Men Referred for Nuclear Stress Testing." Archives of Internal Medicine, Volume 166, Number 2, Pages 201-206. 2006. http://archinte.ama-assn.org/cgi/reprint/166/2/201

Many men who are older than 50 or 60 years of age may helplessly accept erectile dysfunction as a part of the aging process. They do not know erectile dysfunction reflects the health of their cardiovascular and other systems. They may notice erectile dysfunction after they have been diagnosed with diabetes (DM) and/or cardiovascular diseases. They may also have noticed mild erectile dysfunction ahead of the development of diabetes mellitus and major cardiovascular diseases. It is a blessing if the erectile dysfunction appears before other medical disorders are well advanced and it prompts the patient to bring it up to his physician for further tests for their general health. Without dealing with the underlining causes for erectile dysfunction, but going ahead to use chemicals for improving erectile function, it will be a missed opportunity for a health shake-up.

This study reported that, using stress myocardial perfusion single-photon emission computed tomography (MPS), as compared to the non-ED patients, the ED patients had 2.50 times of risk for severe coronary heart disease, and had 2.86 times of chance for positive findings of high-risk MPS. MPS is a Nuclear Stress Test for the cardiovascular system.

Hyperglycemia And Female Sexual Dysfunction

Caruso S, et al. "Sildenafil improves sexual functioning in premenopausal women with type 1 diabetes who are affected by sexual arousal disorder: a double-blind, crossover, placebo-controlled pilot study." Fertility and Sterility. Volume 85, Issue 5, Pages 1496-1501. May 2006. http://www.sciencedirect.com/science?_ob=ArticleURL&_udi=B6T6K-4JKYWGR-2&_user=10&_rdoc=1&_fmt=&_orig=search&_sort=d&view=c&_acct=C000050221&_version=1&_urlVersion=0&_userid=10&md5=1ab79703bdd0c9d481fdcc758574d078

In recent years there has been a good number of studies done on erectile dysfunction (ED) in men, but we have not studied much about sexual dysfunction in women. Women's sexual dysfunction includes loss of sexual desire, unhappiness in sexual relationship, lack of orgasm in sexual relationship, and et cetera. To understand its mechanism, researchers have conducted studies on dysfunction of the woman's clitoris, which is the counterpart of the man's penis. As suspected, women's clitoral dysfunction shares similar findings to those involved in men's penile erectile dysfunction. This pilot study showed that type 1 diabetic women experienced improvement with Sildenafil (the generic name of Viagra) in their sexual dysfunction caused by diabetes mellitus.

Hyperglycemia, CHD, And Female Sexual Dysfunction

Kaya C, et al. "Sexual function in women with coronary heart disease: a preliminary study." International Journal of Impotence Research. Volume 19, Pages 326–329. 2007. http://www.nature.com/ijir/journal/v19/n3/abs/3901530a.html

Hyperglycemia inhibits the production and release of nitric oxide from the endothelium (the inner layer of the blood vessel). Nitric oxide plays an important role in dilating the blood vessels and reducing blood pressure. It is also a primary factor in helping penile or clitoral engorgement.

As reported in cases of man's erectile dysfunction, woman's sexual dysfunction should be carefully evaluated for possible indication of more serious medical problems, such as diabetes (DM), arteriosclerosis, and coronary heart disease (CHD). This study found a close link between female coronary heart disease and sexual dysfunction. The level of sexual dysfunction was apparently more significant among the female patients who had coronary heart disease. Like men, women who experience sexual dysfunction, should have a careful evaluation on their health, including but not limited to, the possibility of developing diabetes (DM), coronary heart disease, and other disorders.

CARBOHYDRATES AND THE FEMALE REPRODUCTIVE SYSTEM

Hyperglycemia (high blood glucose) impacts many, if not all, aspects of our health. To no exception, it affects women's health both during and after their childbearing years.

Most people realize that hyperglycemia during pregnancy (gestational diabetes) presents a high risk of giving birth to a big baby. However, not all the women who gave birth to a big baby were diagnosed with diabetes (DM) before childbirth. These women probably had abnormally high blood glucose levels that had been undiagnosed. (Please see Diabetes Mellitus, Glossary of important Terms, Chapter 1,)

It is interestingly to note that women who became pregnant as diabetics often find their blood glucose levels lower when they are pregnant. This is because that the mother's (high) blood glucose is shared with the fetus, through the placenta; the fetus has helped use the excessive blood glucose. High blood glucose levels can damage ß (beta)-cells of the pancreas, which produces insulin in response to higher blood glucose. *I suspect the newborn's pancreas would be damaged so much that the baby would become diabetic in his infancy or early childhood.*

During recent years, more researches have shown that inflammation is at least partly, if not solely, the cause(s) of numerous diseases affecting the female reproductive system. Many reports have suspected that either local or systemic inflammation, or both are responsible for several gynecological disorders, such as endometriosis, endometrioma, polycystic ovarian syndrome, pelvic inflammatory disease, and et cetera. Unfortunately, these studies were not designed to look into the possibilities of hyperglycemia in these disorders, although studies have found hyperglycemia causes local and systemic inflammation.

Type 1 Diabetes Mellitus And Menstruation

Strotmeyer ES, et al. "Menstrual Cycle Differences Between Women With Type 1 Diabetes and Women Without Diabetes." Diabetes Care, Volume 26, Number 4, Pages 1016-1021. April 2003. http://care.diabetesjournals.org/cgi/reprint/26/4/1016.pdf

Hyperglycemia (high blood glucose) impacts our health, including the endocrine system (hormone) and the reproductive system. This study found that as compared to women who did not have diabetes (DM), the type 1 diabetic women had higher chances of starting their first menstrual (menarche) cycle later in age, early menopause, fewer pregnancies, more stillbirths, and twice higher risk for more problems with their menstruation before they reached 30 years of age. The problems included longer cycles, longer periods, and heavy bleeding.

Chronic Inflammation And Polycystic Ovarian Syndrome

Kelley CCJ, et al. "Low Grade Chronic Inflammation in Women with Polycystic Ovarian Syndrome." Journal of Clinical Endocrinology and Metabolism, Volume 86, Number 6, 2453-2455. 2001. http://jcem.endojournals.org/cgi/content/abstract/86/6/2453

Polycystic ovarian syndrome (PCOS) is a group of disorders for women who have hyperglycemia (abnormally high blood glucose), diabetes (DM), insulin resistance (abnormally high blood glucose, despite a high level of insulin in the circulation), weight gain, thickening of the outmost capsule of the ovary (which is thought to be the reason for preventing ovulation and creating multiple cysts under the capsule), no ovulation, scanty menstruation, infertility (unable to become pregnant), excessive production of testosterone (male hormone), excessive growing of body hairs, acnes, overweight/obesity, high risk of coronary heart disease, and et cetera. The blood test result for chronic low-grade inflammation finds an increase of C-reactive protein (CRP), along with other inflammatory factors.

This article showed that the women who had PCOS had three times higher value of C-reactive protein CRP (in average, 2.12 mg/L vs. 0.67 mg/L). The higher the value of CRP, the lower the insulin sensitivity or the higher the insulin resistance. (Note: Hyperglycemia causes inflammation). This study suggested that, because of the above findings and hyperglycemia, these women with PCOS might have higher risks of coronary heart disease and diabetes mellitus.

Low Carbohydrate Diet For Polycystic Ovarian Syndrome

Mavropoulos JC, Yancy WS, Hepburn J and Westman EC. "The effects of a low-carbohydrate, ketogenic diet on the polycystic ovary syndrome: A pilot study." Nutrition and Metabolism, Volume 2, Number 35. 2005. http://www.nutritionandmetabolism.com/content/2/1/35

Polycystic Ovarian Syndrome (PCOS) happens in about one in ten women. PCOS is characterized by the symptoms described in the previous article. This study investigated the impacts of low-carbohydrate, ketogenic diets (LCKD) on patients who had PCOS, for six months. The LCKD for this study contained 20 grams or less of carbohydrates a day. The diet helped women who had PCOS to lose weight. The diet lowered free testosterone (male sex hormone), radically decreased insulin in the blood circulation, and improved the degree of insulin sensitivity. Interestingly, in this study, two women with a history of infertility became pregnant.

CARBOHYDRATES AND CHILDHOOD

Childhood obesity has become an epidemic in the United States and many other parts of the world. The data from National Health and Nutrition Examination Survey (NHANES), during the years from 1963 and 1970, showed that only 4% of the children between 6 and 11 years old and 5% of the children between 12 and 19 years old, were overweight. During the period between 1976 and 1980, 7% of the 6-11 years old and 5% of the 12-19 years old were overweight. However, the period between 1999 and 2002, 16% of the 6-11 years old and 16% of the 12-19 years old were overweight. In the most recent studies, 17.1 % of the children were overweight or obese.

We try to make sure that our children are safe inside a car, by asking the car seat manufacturers to design larger car seats for overweight children. The most important work that we all must do is to find the right way to help children lose weight.

Unfortunately, most physicians and nutritionists still fail to recognize that they have given the public incorrect advice as to what a healthy diet should include. For example, removing French fries from the school lunch is an excellent move. However, doing so in an effort to help children avoid fried foods is the wrong reason. We should avoid French fries because they are starchy food. Eating too many carbohydrates makes children want to eat more and gain weight. In addition, frying carbohydrates with oil or fat, particularly at such a high temperature raises the risk of glycation. The endproduct of advanced glycation is toxic (AGE) to our body. (See Glycation in this chapter.)

Obesity is a serious health problem. More overweight or obese children are suffering more diseases, which were formerly found almost solely in their adult counterparts. J. S. Olshansky and coworkers published a research report, in the March 17, 2005 issue of New England Journal of Medicine. They were concerned that, with an increasing chance of having the adulthood diseases, these overweight or obese children are expected to shorten their lives. We must correctly identify the cause(s) of excessive weight gain in children, and effectively reverse the trend of obesity in the population, especially among children.

Gestational Diabetes And Congenital Malformation

Narchi H, et al. "High incidence of Down's syndrome in infants of diabetic mothers." Archives of Disease in Childhood. Volume 77, Pages 242–244. 1997. http://www.pubmedcentral.nih.gov/picrender. fcgi?artid=1717320&blobtype=pdf

We have newborns everyday. We do not realize how lucky we are to be able to welcome a normal baby, until we have one with some sort of malformation. Many studies have already linked congenital malformation and birth defects to mothers who had diabetes (DM) before or during their pregnancies.

This study found that diabetic mothers have a 2.75 times higher chance of delivering a baby with Down's syndrome than non-diabetic mothers. This study also found no significance in varying ages of the mothers at the time of their pregnancies. We must keep it in mind that non-diabetic pregnant women could have high blood glucose after meal that increases the risk of congenital malformation of their babies.

Gestational Diabetes And Child Obesity

Vohr BR, McGarvey ST and Tucker R. "Effects of Maternal Gestational Diabetes on Offspring Adiposity at 4–7 Years of Age." Diabetic Care, Volume 22, Number 8, Pages 1284-1291, 1999. http://care.diabetesjournals.org/cgi/reprint/22/8/1284

This study observed that children who were born to diabetic and obese mothers grew fatter than other children during the period between the ages of four and seven years. (Note: Since children usually eat at the same table with obese parents, is not it possible that the children grow fatter because of the similarity in their dietary habit to that of their parents? We should not blame genetics for everything, before looking into other factors!)

Childhood Obesity And Car Seat Safety

Trifiletti LB, et al. "Tipping the Scales: Obese Children and Child Safety Seats." Pediatrics. Volume 117, Number 4, Pages 1197-1202. 2006. http://pediatrics.aappublications.org/cgi/reprint/117/4/1197

This article reported the growing problem of having difficulties in finding appropriate car seats for children. It found 283,305 children between 1 and 6 years of age, who were overweight or obese at the time of its study. It urged everyone who is involved in providing a safe child car seat to deal with child obesity seriously, and to give the obese children safe car seats fitting the child's body size.

Childhood Obesity And Adulthood Health Disorders

Vanhala M, et al. "Relation between obesity from childhood to adulthood and the metabolic syndrome: population based study." British Medical Journal, Volume 317, Number 7154, Pages 319–320. August 1, 1998. http://www.pubmedcentral.nih.gov/picrender. fcgi?artid=28624&blobtype=pdf

Childhood obesity has a strong link to adult obesity. Adult obesity is often linked to many diseases, particularly metabolic syndrome. Metabolic syndrome is a group of medical disorders including high blood pressure, abnormal blood lipid (fat) profile, such as an increase of low-density lipoprotein-cholesterol (LDL, especially VLDL) and a decrease of high-density lipoprotein-cholesterol (HDL), and high blood glucose, particularly when the amount of circulating insulin is also increased. The latter situation indicates that the person not only has diabetes (DM), but also has a serious problem, that his body cannot handle the excessive blood glucose. This research report found a strong link between an increased risk of metabolic syndrome and the continued obesity from one's childhood into adulthood.

Low Carbohydrate Diet And Child Obesity

Sondike SB, Copperman N and Jacobson MS. "Effects of a low-carbohydrate diet on weight loss and cardiovascular risk factor in overweight adolescents." Journal of Pediatrics, Volume 142, Issue 3, Pages 253-258. March 2003. http:// pediatrics.aappublications.org

The most common causes linked to the obesity epidemic

are (1) taking in too many more calories than our body needs, while most often blaming fat for the problem; (2) lack of physical activities, such as little exercise and too much time in watching television and playing computer games, riding more than walking for a short distance; (3) genetic influence; (4) environmental and cultural; (5) psychological. It is understandable that eating too many calories is a most important factor in gaining weight but it is doubtful whether fats are the cause, after further studying as to how carbohydrates affect the consumption and storage of energy in the body.

This study looked into the effects on weight loss and serum lipid profile of a low-carbohydrate diet, for a small group of overweight adolescents. The study found the overweight teenagers, who followed a low-carbohydrate diet, lost twice as much weight as those who followed a low-fat diet during the study period. The study concluded that the low-carbohydrate diet was effective in reducing the overweight adolescents' body weight, with no harm to their lipid profiles for the short term.

CARBOHYDRATES AND THE NERVOUS SYSTEM

Recently some physicians have begun using the phrase "brain attack" instead of "stroke," as they similarly use the phrase "heart attack" for acute ischemic coronary heart disease. However, both stroke and heart attack are laymen's terms; they do not clearly identify the nature of the diseases.

Stroke is brain damage as a result of vascular disorders. There are two kinds of stroke. One is ischemic (lack of blood circulation) and the other is hemorrhagic (bleeding). In an ischemic stroke, arteries become clogged by blood clot(s) or debris. This stops the blood circulation to the vital area of the brain. Usually, the ischemic stroke is a result of arteriosclerosis or atherosclerosis. The hemorrhagic stroke occurs when a blood vessel(s) is ruptured and bleeds into the brain tissues. Except for cases of trauma or a bleeding tumor, this is usually a result of rupture of an aneurysm. An aneurysm is the weakened wall of the blood vessel, which has been damaged by the development of atherosclerosis.

Hyperglycemia (high blood sugar) is the most dangerous player in developing atherosclerosis. Other than in diabetic conditions, hyperglycemia usually happens after eating a high-carbohydrate meal.

More research reports, during recent years, have found that blood glucose played an important role in causing neurological and mental disorders, such as multiple sclerosis, amyotrophic lateral sclerosis, Alzheimer's disease and dementia, and Parkinson's diseases. These reports are included in the section, "Carbohydrate and Glycation."

High Blood Sugar And Stroke

Williams LS, et al. "Effects of admission hyperglycemia on mortality and costs in acute ischemic stroke." Neurology Volume 59, Pages 67-71. 2002. http://www.neurology.org/cgi/content/abstract/59/1/67

Based on recent studies, we have come to understand that controlling the blood glucose (sugar) level, in an emergency or medically critical situation, importantly affects the outcome of the situation. Hyperglycemia can trigger inflammation and further damage already affected tissue(s).

This study investigated the link between the blood glucose level of a stroke patient, at the time of admission to the hospital, and his prognosis with acute ischemic stroke. In this study, the cut-off point for high blood glucose level was defined as a level of 130 mg% or over. The upper limit of normal fasting blood glucose is 100 mg%. The study showed 40% of the patients had high blood glucose level at their admission to the hospital with acute ischemic stroke. The study also showed longer hospitalization and higher costs for the care of patients with high blood glucose. These patients had higher risks for death at 30 days (1.87 times), one year (1.75 times), and six years (1.41 times), after the onset ischemic stroke.

Neurological Disorders And Inflammation

Ridker P and Samuel MA. "Inflammation and Neurological Disease." Continuum, Volume 11, Issue 1, Pages 114-118. February 2005. http://www.aan.com/elibrary/continuum/?event=home.showIssue&issue=ovid.com:/issue/ovftdb/00132979-200502000-00000

Up to this date, we still do not know much about the causes, of multiple sclerosis (MS), amyotrophic lateral sclerosis (ALS or Lou Gehrig's disease), Guillain Barré syndrome, and many other diseases impairing the nervous system. We have suspected viruses, bacteria, autoimmune disorders, and others, as possible causes for the diseases. However, are they really the prime suspects?

We know that all of the above suspects can cause infection or destruction to body tissues. Advancement in molecular and cellular chemistry has helped us discover that neurological diseases are also closely related to inflammation. Inflammation is one thing common to all diseases, including the neurological ones.

This study cited that many neurological diseases, such as meningitis and encephalitis (brain infection), some of the known neuropathies (damages to the nerves such as diabetic neuropathy), and multiple sclerosis, are a result of inflammation. The study suggested inflammation was a systemic phenomenon, involving the nervous system, through the inflammatory reflex. Many neurological diseases, such as migraine and Alzheimer's disease may involve inflammation in their disease process. Controlling the low-level inflammation could be a therapeutic goal, to slow down the development of many neurological diseases.

Hyperglycemia And Retinopathy

Brinchmann-Hansen O, et al. "Blood glucose concentrations and progression of diabetic retinopathy: the seven year results of the Oslo study." British Medical Journal, Volume 304, Number 6818, Pages 19–22. January 4, 1992. http://www.pubmedcentral.nih.gov/articlerender.fcgi?artid=1880908

It is painful to lose one's eyesight. It would be even more painful to realize that one could have prevented the loss if he had altered his dietary choices to avoid the rise of his blood sugar. Many of us have been told by our doctors that diabetes (DM) can damage our eye ground (retina), which is connected to the optic nerve (the nerve for eyesight). Damage to the retina (retinopathy) can include build-up of microaneurysm (aneurysm of a small artery, which is a result of atherosclerosis) and hemorrhage (bleeding from the weakened and ruptured microaneurysm).

This article found that intensified insulin treatment helped bring down the blood glucose level, reflected in hemoglobin A_{1c}, and prevented the advance of retinopathy. In the June 12, 2005 issue of NIH News (National Institute of Health http://www.nih.gov/news/pr/jun2005/niddk-12. htm), it was reported that nearly 8% of the participants in the Diabetes Prevention Program (DDP), who were pre-diabetic and had already been found to have retinopathy. This report alerts us that retinopathy happens much earlier than we have thought.

Hyperglycemia And Head Injury

Aristedi R and Serafim K. "The Influence of Hyperglycemia on Neurological Outcome in Patients with Severe Head Injury." Neurosurgery, Volume 46, Number 2, Pages 335, February 2000. http://www.neurosurgery-online. com/pt/re/neurosurg/abstract.00006123-200002000-00015.h tm;jsessionid=HbJb6gYJnLlcwYrKt3C4W8s2V8Y8v2zGtN yfQspfwrPx1CDZcP01!-1288052477!181195628!8091!-1

Injury is a terrible thing that no one wants to experience. It is even worse when someone you love has a head injury and lies unconscious, yet the doctor can offer no encouraging prognosis. The doctor cannot predict complications such as ischemia (lack of blood circulation) and hypoxia (short of oxygen supply) to the injured brain tissue. During recent years, medical researchers have found hyperglycemia (high blood glucose) has an important role in deciding the fate of the injured. The results of the study showed that the blood glucose level was higher in patients who had suffered severe head injuries than those who had incurred only moderate injuries. The patients who had a poor outcome had higher blood glucose levels than those who had a better outcome. The patients who had blood glucose levels higher than 200 mg% had a more perilous outcome. The study concluded that hyperglycemia in the early stages after a head injury reflected the stress response to that injury was a significant indicator of its severity as well as a reliable predictor of outcome.

CARBOHYDRATES AND GLYCATION

Glycation is a bonding reaction in which a sugar molecule bonds itself to a protein or fat (lipid) molecule, or both, either with or without the involvement of enzyme. The typical example for the enzymatic glycation or glycosylation (with the involvement of enzyme) is the bonding between blood glucose (sugar) and globin (the protein part) of hemoglobin. Hemoglobin is responsible for transporting oxygen and carbon dioxide in the red blood cells between the tissues and the lungs. The higher the blood glucose level, the higher percentage point of the hemoglobin A_{1c} will be. We use the reading of hemoglobin A_{1c} to detect the level of blood glucose, especially in case of diabetes (DM). (Please see Hemoglobin A_{1c}, Glossary of Important Terms, Chapter 1.) The other example of glycosylation in this book is the bonding between osteocalcin and glucose. Free osteocalcin in the blood circulation is decreased in diabetic patients. As a result, the diabetic patient has a higher risk of fractures.

The non-enzymatic glycation includes exogenous (outside the body) and endogenous (inside the body) glycation. The product of glycation is simply called glycation product or glycoprotein and glycolipid, respectively. The sugar (or carbohydrate) portion of these glycated products is called glycan. Glycation can grow, with the glycation products, and produce advanced glycation endproducts

301

(AGEs).

A typical exogenous glycation takes place in the cooking of sugar together with either protein or fat, or both. Heat speeds up glycation. However, lower heat in a prolonged cooking of sugar with protein or fat can produce advanced glycation endproducts too. Therefore, prolonged cooking can increase glycation. About 30% of the exogenous AGEs are absorbed through digestion, which are pro-inflammatory (promoting inflammation) and toxic to the body. They are carcinogenic (cancer-causing). They may also cause diabetes mellitus, cardiovascular diseases, retinal dysfunction, and many other diseases, probably because of their ability to promote inflammation. AGEs are plentiful in food flavor and color additives. We add sugar to products in cooking for browning effect, such as French fries and baked foods.

Endogenous glycation involves fructose, galactose, and glucose inside the body. Fructose and galactose are about ten times more active than glucose in glycation. Endogenous AGEs are implicated in diseases such as diabetes mellitus, cardiovascular diseases, cancers, Alzheimer's disease (amyloid-β), and many others. Moreover, glycation can also cause genetic mutation by bonding the protein of DNA with a sugar molecule. Glycation on DNA changes the genetic codes. Mutation can possibly cause allergies, diabetes (DM), cancers, and congenital malformations.

Dietary Glycotoxin Induces Inflammatory Mediators

Vlassara H, et al. "Inflammatory mediators are induced by dietary glycotoxins, a major risk factor for diabetic angiopathy." Proceedings of the National Academy of

Science of the United States of America, Volume 99, Number 24, Pages 15596–15601. November 26, 2002. http://www. pubmedcentral.nih.gov/articlerender.fcgi?tool=pmcentrez&a rtid=137762

AGEs (advanced glycation end-products) are both exogenous and endogenous, and recognized as a cause of inflammation in the body. This study observed that by cooking the same foods at different temperatures and for different length of time, one diet could contain more of AGEs than the other. The results clearly demonstrated that diets with high AGEs caused a higher possibility of inflammation and vascular complications. The study found that avoiding high AGE diets would prevent or decrease inflammation and damages to the blood vessels, such as atherosclerosis in diabetic patients.

Glycemic Control Improves Skin Structure

Lyons TJ, et al. "Decrease in Skin Collagen Glycation with Improved Glycemic Control in Patients with Insulin-dependent Diabetes Mellitus." Journal of Clinical Investigation, Volume 87, Number 6, Pages 1910–1915. June 1991. http://www.pubmedcentral.nih.gov/articlerender. fcgi?tool=pmcentrez&artid=296942

Collagen is a protein and an important component of connective tissue of animals, such as vascular wall and skin. Protein is a chain of peptides, which are formed by more than one amino acid. Collagen has lysine, an amino acid at the free end of the protein. Therefore, collagen can be the target for glycation, and thus damaged by it. This study investigated the impact of hyperglycemia on skin collagen. Intensive glycemic control for lowering their blood glucose reduced both the individual's hemoglobin A_{1c} and the rate of glycation of the collagen fibers. As a result, the condition of the individual's skin was significantly improved.

Glycation And Lou Gehrig's Disease

Takamiya R, et al. "Glycation proceeds faster in mutated Cu, Zn-superoxide dismutases related to familial amyotrophic lateral sclerosis." FASEB Journal, Express Article, 10.1096/fj.02-0768fje. http://www.fasebj.org/cgi/content/abstract/02-0768fjev1

Amyotrophic lateral sclerosis (ALS) is also called Lou Gehrig's disease after the famous American baseball player who died from it. It is a gradual weakening of the motor nerve cells of the spinal cord and the motor segment of the brain. The ALS patient will gradually lose the use all of his muscles before death. However, ALS does not involve his mental capacity. There are two types of ALS, the familial type with history involving the heredity (FALS) and the sporadic type with no family history of the disease. The familial ALS is usually found a patient's defective genes that are thought to cause the disease. *Are not the defective genes themselves a possible result of glycation and hyperglycemia?*

As of this date, it is generally agreed that oxidation is one of the causes for damage and degeneration of tissues. That is the reason why many of us are taking different antioxidants, in hope of decreasing the level of oxidation in our body. To no exception, ALS may be partly caused by oxidation. This study found that the defective enzyme (SOD) produced under the message from the defective gene of the FALS patients was much likely involved in glycation, or fructation (fructose bonded with the protein of the enzyme) in this case. The glycated SOD (enzyme) tended to produce more hydrogen peroxide, which could increase oxidation (or oxidative stress in medical terminology) with the nerve cells in FALS patients. As a result of oxidation, the nerve cells were damaged and became dysfunctional.

Glycation And Parkinson's Disease

Münch G, et al. "Crosslinking of α-synuclein by advanced glycation endproducts — an early pathophysiological step in Lewy body formation." Journal of Chemical Neuroanatomy, Volume 20, Issues 3-4, Pages 253-257. December 2000, http://www. sciencedirect.com/science?_ob=ArticleURL&_udi=B6T02-42C07TD-6&_user=10&_coverDate=12%2F31%2F2000&_rdoc=1&_fmt=&_orig=search&_sort=d&view=c&_acct=C000050221&_version=1&_urlVersion=0&_userid=1 0&md5=79c1965fdbd0c91e20bbb4f8aa1537ac

Parkinson's disease (Parkinson Disease or PD) affects one's motor functions, as a result of degenerative changes of the central nervous system. The basal ganglia (pleural of ganglion. Dense clusters of nerve cells) of the large brain (cerebra) send less and less stimulation (signal) to the motor cortex (the region of the large brain controls motor functions).

The causes of degenerative changes in the nervous system, such as Parkinson disease, include advanced glycation endproducts (AGEs), among other things. Parkinson disease is characterized by a build-up of Lewy bodies, which are AGEs and thickly formed protein deposits inside the nerve cells of the PD patient's brain (mainly, the substancia nigra or black substance). These insoluble Lewy bodies are glycated alpha synuclein, which is soluble before glycation. Lewy bodies have been found in the late stage of Parkinson disease. This study suggested that glycation of soluble alpha-synuclein in forming insoluble Lewy bodies was the cause of degenerative changes in the early stages of Parkinson disease.

Glycation And Multiple Sclerosis

Kalousova M, et al. "Advanced Glycoxidation End Products in Patients with Multiple Sclerosis." Prague Medical Report, Volume 106, Number 2, Pages 167–174. 2005. http://www.galenicom.com/pt/medline/article/ 16315765

Several neurological diseases, including amyotrophic lateral sclerosis, Alzheimer's disease, Parkinson disease, and multiple sclerosis, are a result of glycation and its related biochemical products including the receptor for AGEs (advanced glycation endproducts) or RAGE.

This study found that the amount of AGEs did not provide a good indicator for the risk of multiple sclerosis. Rather, these AGEs molecules and the molecules of the receptor of AGEs (RAGE) were probably the more important factors in the disease. Another study, in the Columbia University health Science online publication, In Vivo, Volume 2, No. 5, published on March 12, 2003 (http://www. cumc.columbia.edu/news/in-vivo/Vol2_Iss05_mar12_03/ index.html), found that jamming the RAGE activity, into the immune cells in a mice model, could put off a majority of multiple sclerosis-like symptoms.

CARBOHYDRATES AND VISION

Carbohydrates And Cataracts (1)

Chiu CJ, et al. "Carbohydrate intake and glycemic index in relation to the odds of early cortical and nuclear lens." opacities. American Journal of Clinical Nutrition, Volume 81, Number 6, Pages 1411– 6. 2005. http://www.ajcn.org/cgi/reprint/81/6/1411.pdf

A normal lens of the eye is as clear as crystalline. A cataract is diagnosed when the lens becomes cloudy. A cataract is one of the most common diabetic complications. Depending on the locations of the lens opacity, there are anterior (frontal) cortical cataracts, anterior subcapsular (underneath a capsule) cataracts, anterior polar cataracts, nuclear cataracts, posterior (rear) cortical cataracts, posterior subcapsular cataracts, and posterior polar cataracts. The term "nuclear" refers to those near the center of the lens. The term "cortex" describes those which surround the nuclear.

This study grouped individuals based on the amount of total daily carbohydrate consumption (less than 185, 185-200, and more than 200 grams/day). The study found that the highest group in the category of daily carbohydrate consumption had 2.46 times higher risk of developing cortical cataracts than those in the lowest group.

Carbohydrates And Cataracts (2)

Chiu CJ, Milton RC, Gensler G and Taylor A. "Dietary carbohydrate intake and glycemic index in relation to cortical and nuclear lens opacities in the Age-Related Eye Disease Study." American Journal of Clinical Nutrition, Volume 83, Number 5, 1177-1184, May 2006. http://www.ajcn.org/cgi/

content/abstract/83/5/1177

This study grouped individuals based on the amount of daily carbohydrate consumption and on the glycemic index of the foods they ate. The study observed that dietary glycemic index and glycemic load might be links to both the nuclear and cortical cataracts. Please see the classification of cataracts in the previous article, CARBOHYDRATES AND CATARACTS (1).

Carbohydrates And Glaucoma

Oshitari T, Fujimoto N, Hanawa K and Adachi-Usami E. "Effect of Chronic Hyperglycemia on Intraocular Pressure in Patients With Diabetes." American Journal of Ophthalmology, Volume 143, Issue 2, February 2007. http://www.sciencedirect.com/science?_ob=ArticleURL&_udi=B6VK5-4M4KJ1D-4&_user=10&_coverDate=02%2F28%2F2007&_rdoc=1&_fmt=&_orig=search&_sort=d&view=c&_acct=C000050221&_version=1&_urlVersion=0&_userid=10&md5=0bbe736b12fc743657ee46c535e3f88f#bcor1

Glaucoma is a disease caused by an increase of pressure inside the eyeball that causes pain and loss of eyesight. The many risk factors for glaucoma include a family history of the disease and diabetes (DM). African Americans are more likely to have this disease than others. This study observed that the patients with chronic severe hyperglycemia (diabetes mellitus) had significantly and progressively higher readings of intraocular pressure (glaucoma) than those with mild hyperglycemia. Controlling the blood glucose level is very helpful!

Carbohydrates And Macular Degeneration

Chiu CJ, et al. "Dietary glycemic index and carbohydrate in relation to early age-related macular degeneration." American Journal of Clinical Nutrition, Volume 83, Number 4, 880-886, April 2006. http://www.ajcn.org/cgi/content/full/83/4/880

Macular degeneration is most commonly found in older adults. The macula is the center of the retina or eye-ground. When it is degenerated, it gets thinner and it may shrink in size or bleed. The patient with macular degeneration loses central vision, and cannot read or recognize things. Many risk factors have been mentioned, including smoking, family history and heredity or gene, hypertension, dietary fats, cholesterol, race, exposure to sunlight, and et cetera. We have seldom mentioned dietary carbohydrates.

This study categorized individuals into three groups based on the glycemic index of the foods they consumed, 74.6% and below, 74.6-77.0%, and over 77.0%, respectively. The individuals with the highest glycemic index of foods had 2.71 times higher risk for macular degeneration in related to the individuals with the lowest GI. *A glycemic index at 74.6% is already too high!*

CARBOHYDRATES AND DEMENTIA

Midlife Obesity And Old Age Dementia

Rosengren A, Skoog I, Gustafson D and Wilhelmsen L. "Body Mass Index, Other Cardiovascular Risk Factors, and Hospitalization for Dementia." Archives of Internal Medicine, Volume 165, Number 3, Pages 321-326. FEB 14, 2005. http://archinte.ama-assn.org/cgi/content/

abstract/165/3/321

This study reviewed the link between body mass index (BMI) and the risk of dementia. The study observed that the higher the individual's body mass index, the higher the risk of dementia (2.45 times for a BMI over 30). The study used individuals with a BMI at 20-22.49 as the baseline group with 1.0 for the smallest risk of dementia. However, the risk of dementia is 2.38 times higher for the individuals with a BMI below 20.

It is interesting to note that a very high risk of dementia exists in the group of individuals who had the smallest BMI. The result supports the notion that hyperglycemia, not necessarily BMI, is responsible for dementia. Hyperglycemia does not always cause obesity. One could consume an amount of daily calories equal to the amount of calories for his daily activities, and yet remain underweight. However, he could have hyperglycemia after each meal if he gets most of his calories from carbohydrates.

Diabetes Mellitus And Dementia Or Alzheimer's Disease

Leibson, CL, et al. "The Risk of Dementia among Persons with Diabetes Mellitus: A Population-Based Cohort Study. American Journal of Epidemiology, Volume 145, Number 4, Pages 301-308. 1997. http://aje.oxfordjournals. org/cgi/reprint/145/4/301

Over the years, we have continued to look for the cause(s) for Alzheimer's disease and dementia. Recent research has found a relationship between diabetes (DM); and Alzheimer's disease; and dementia. This population-based study found that the adult-onset diabetic patients had

1.66 times more for the risk of all dementia than the people who were not diabetic. The risk of Alzheimer's disease was 2.27 times for men and 1.37 times for women, respectively, who were adult-onset diabetic, as compared to people who were not diabetic.

Type 2 Diabetes, Glycation, And Alzheimer's Disease

Janson J, et al. "Increased Risk of Type 2 Diabetes in Alzheimer Disease." Diabetes, Volume 53, Number 2, Pages 474-481. 2004, http://diabetes.diabetesjournals.org/cgi/content/abstract/53/2/474

Amyloid is a glass-like biomolecule. "Amyl-" means "starch." "-oid" means "like something." So, amyloid was thought to be something like starch. After it is chemically stained with a dye, Congo Red, it looks a smooth, pink or red glass. ß (beta)-amyloid is an insoluble (non-dissolvable), hard protein structure, converted from a soluble (dissolvable), soft protein amyloid, by the process of glycation. (Please see CARBOHYDRATES AND GLYCATIONS in this chapter.) In an article, "Glycation Stimulates Amyloid Formation", Mark B. Obrenovich and Vincent M. Monnier of Department of Pathology, Case Western University pointed out that evidence supports the involvement of glycation, in forming ß (beta)-amyloid. Amyloid can be found both inside and outside of a cell. (Glycation Simulates Amyloid Formation, Science of Aging Knowledge Environment, Vol. 2004, Issue 2, pp. pe3, 14 January 2004)

The amyloid structure can further involve itself in glycation and grow bigger and bigger, like a snowballing effect. When glycation occurs outside a cell, amyloid becomes a plague, and can also grow and damage the cellular structure and functions.

Research on Alzheimer's disease and diabetes mellitus during the recent years has found amyloid in both the brain tissue of the patients with Alzheimer's disease, and in the ß-islet cells of the pancreas of the patients with diabetes mellitus. In case of Alzheimer's disease, insoluble beta-amyloid is derived from soluble amyloid. In case of type 2 diabetes mellitus (DM), amyloid is derived from amyloid polypeptide. In addition, they share similar circumstances, such as increasing risks of the diseases in the older population and a genetic connection.

This study reviewed the relationship between Alzheimer's disease and type 2 diabetes (DM), based in both clinical and pathological reviews. Clinically, a majority of the Alzheimer's disease (81%) had either diabetes (DM) or abnormal or high fasting blood glucose. On the other hand, the study found a strong positive relationship in the autopsy findings of amyloid formation between the brain and the pancreas. The study concluded that Alzheimer's disease and diabetes mellitus are closely linked. *Based on this study as well as others, preventing hyperglycemia at anytime is the key(s) to avoiding glycation, amyloid formation, and other disease processes. Since the level of blood glucose is directly tied the amount and the grade or quality of the dietary carbohydrates, does not it make more sense to avoid consuming an excessive amount of carbohydrates?*

CARBOHYDRATES AND CANCERS

Before learning how to plant grains and vegetables, man had used meat as his main source of nutrition. He used green leaves, seeds, fruits, and wild plants for supplement. As civilization developed, man began to replace more meat with agricultural products, such as grains, produce, fruits, and others. With technological advancement in improving the quality and availability of the agricultural products, man became very fond of sweet products, especially sugar, grains, grain products, and fruits.

For the past decades, thanks to the national policies of the US and other countries for protecting their agricultural production, Americans and people of other countries have consumed more grains and fruits, which are abundant in carbohydrates.

During the same period, nutritionists and physicians have believed that carbohydrates are the best source of nutrition. They realize that carbohydrates generate only 4 kcal/grams of energy, while fats generate 9 kcal/grams, and proteins 4 kcal/grams. Based on the individual's daily energy requirement, people can afford to take in more carbohydrates in volume than fats. In addition, both nutritionists and physicians believe that fats are the cause of obesity, cardiovascular diseases, stroke, cancers, and other disorders. They have become "lipophobic." At the same time, they are still unsure of the impacts on our health from using more proteins for nutrition. They are somewhat "aminophobic." (I made up the word that means "afraid of protein.") With all of the above, nutritionists and physicians continue to promote more consumption of carbohydrates than fats and proteins. Also, the latest food pyramid still recommends heavy consumption of carbohydrates.

We have come across a rapid rise in the number of serious medical problems during the recent decades, especially diabetes mellitus (DM), and complications as the result of obesity. In the meantime, we have also witnessed a growing number of cancers, for which, until the recent years, we had thought fats were responsible. Now, more evidence shows a link between hyperglycemia and some cancers. With better design for research projects, we are sure that the link is going to become apparent. Most importantly, we must recognize that excessive carbohydrate intake causes hyperglycemia, diabetes (DM), and excess weight or obesity.

While we are still unsure as to the reason why cells mutate themselves into cancer cells, we blame genes, inflammation, chemical reactions, physical irritation, infections, environmental factors, fats, and many others for the development of cancer. As mentioned, we had never suspected carbohydrate consumption as a possible cause until recent years. We do not have to be "carbophobic." (I made up the word that means "afraid of carbohydrates.") We just have to mindfully use carbohydrates. As many more research articles about carbohydrates and cancers are published, I will share with you articles that implicate carbohydrates as the culprit responsible for cancer.

Hyperglycemia And Risks Of Cancers

Jee SH, et al. "Fasting serum glucose level and cancer risk in Korean men and women." Journal of American Medical Association, (JAMA). Volume 293, Number 2, Pages 194-202. January 12, 2005. http://jama.ama-assn.org/cgi/content/abstract/293/2/194?etoc

This study observed the highest death rates from all cancers was related to level of fasting blood sugar (FBS) of

140 mg% and over, and the lowest death rates was related to the FBS at 90 mg% or below. Cancers of the pancreas had the highest link to high fasting blood sugars. It was interesting to note that most of the individuals in this study had a normal body mass index (BMI). Although obesity is on the list of many diseases, hyperglycemia, not obesity, is the cause of the diseases.

High Blood Sugar And Cancer

Stattin P, et al. "Prospective Study of Hyperglycemia and Cancer Risk." Diabetes Care. Volume 30, page 561-567. 2007 http://care.diabetesjournals.org/cgi/content/abstract/30/3/561?ijkey=ef5c7a5dd4b2ae780d2e1c36b26818d5b9dc44be&keytype2=tf_ipsecsha

We are paying more attention on the relationship between hyperglycemia (high blood glucose) and the risk of cancers. More studies have affirmed our concerns. This study investigated the link between fasting hyperglycemia, post-load hyperglycemia, and the risk of cancers of different sites. Post-load hyperglycemia is high blood glucose at the two-hour mark after a glucose tolerance test. The study observed that hyperglycemia was linked to higher risks of cancers in certain sites for both men and women. The risks had nothing to do with the individual's body mass indices (BMI).

Carbohydrate-Rich Diets Cause Stomach Cancer

Augustin LSA, et al.. "Glycemic Index, Glycemic Load, and Risk of Gastric Cancer." Annals of Oncology, Volume 15, Number 4, Pages 581-584. 2004. http://annonc.oxfordjournals.org/cgi/content/abstract/15/4/581

This study investigated the link between an individual's

daily glycemic index (GI) and glycemic load (GL) and the risk of stomach cancer. The study observed that the daily glycemic load, not the glycemic index, was linked to the risk of gastric (stomach) cancers. The study suggested that the overproduction of insulin in the blood, in response to the carbohydrate-rich diets, might be related to the increase of an insulin-like growth factor that might be responsible for the risk of gastric (stomach) cancer.

Diabetes Mellitus And Pancreatic Cancer

Wang F, Herrington M, Larsson J and Permert J. "The relationship between diabetes and pancreatic cancer." Molecular Cancer, Volume 2, Number 4. 2003. http://www.molecular-cancer.com/content/2/1/4

This study cited that 80% of the patients who suffer pancreatic cancer demonstrated glucose intolerance and/or diabetes. However, the study could not be sure whether pancreatic cancer caused diabetes or vice versa.

High Glycemic Load And Pancreatic Cancer

Michaud DS, et al. "Dietary Sugar, Glycemic Load, and Pancreatic Cancer Risk in a Prospective Study." Journal of National Cancer Institute, Volume 94, Number 17, Pages 1293–1300. 2002. http://jnci.oxfordjournals.org/cgi/reprint/94/17/1293

The pancreas produces insulin for regulating blood glucose, keeping it low and within the "normal" range. In the meantime, like other organs and tissues of our body, the level of blood glucose also impacts the well being of the pancreas. Based on other studies, higher blood glucose levels could damage the ß-cells of the pancreas that would decrease or stop the production and release of insulin. There is no doubt that the main source of blood glucose is dietary carbohydrates. The more carbohydrates (glycemic load or GL) consumed, particularly with the refined ones (high glycemic index), the higher the blood glucose level. After eating carbohydrates, one unit of glycemic load can provide one gram of glucose to the blood circulation.

This study found that individuals who are overweight or obese, and get little physical activity, with 148 units of daily GL, had 2.67 times of the risk of pancreatic cancer, as compared to those the lowest daily GL at 93 units. In the meantime, those with the highest consumption of fructose (35 grams) had 3.17 times of risk of pancreatic cancer as compared to those with the lowest daily consumption of fructose (1 gram). Studies also blamed the rise of fructose consumption for the increase in obesity among the US population since the middle of the 20th century.

Obesity And Endometrial Cancer

Kaaks R, Lukanova A and Kurzer MS. "Obesity, Endogenous Hormones, and Endometrial Cancer Risk." Cancer Epidemiology Biomarkers & Prevention Volume 11, Number 12, Pages 1531-1543, December 2002. http://cebp. aacrjournals.org/cgi/content/full/11/12/1531

More evidence proves the link between obesity and the consumption of carbohydrate-rich diets. The latter is the most important cause of hyperinsulinemia. This study observed links to an increased risk of endometrial cancer as well as changes in sexual hormones, obesity, physical inactivity, and hyperinsulinemia (high blood concentration of insulin).

Glycemic Index, Glycemic Load, And Endometrial Cancer

Augustin LSA, et al. "Glycemic Index and Glycemic Load in Endometrial Cancer." International Journal of Cancer Volume 105, Number 3, Pages 404-7. June 20, 2003. http://www3.interscience.wiley.com/cgi-bin/abstract/104081808/ABSTRACT?CRETRY=1&SRETRY=0

This study reviewed data of women from both Italy and Switzerland for the risk of endometrial cancer as related to both the glycemic index and glycemic load. Because of difference in diets between the two countries, grouping the patients based on the glycemic indices (%) and the glycemic load (unit) of their foods was different. As to the glycemic index, the lowest group was less than 78.8 for the Swiss and less than 69.8 for the Italians. The highest group was more than 87.8 for Swiss and more than 80.5 for Italian. As to the glycemic load, the lowest group was less than 108.4 units for the Swiss and less than 84.2 units for the Italians; and the highest group were 213.8 units for the Swiss; and 169.1 units for the Italians.

In comparing to the lowest groups, the risks of endometrial cancer for the highest groups were 2.1 times higher, based on the glycemic indices, and 2.7 times higher, based on the glycemic loads, respectively. The studies also found that the links were even stronger in older women who had heavier body weight and those who had undergone hormone replacement therapy.

Glycemic Index, Glycemic Load, And Ovarian Cancer

Augustin LSA, et al. "Dietary Glycemic Index, Glycemic

Load and Ovarian Cancer Risk: a case–control study in Italy." Annals of Oncology, Volume 14, Number 1, Pages 78-84, 2003. http://annonc.oxfordjournals.org/cgi/content/full/14/1/78

This study divided individuals into groups, based the glycemic indices (%) and the glycemic load of their foods. As to the glycemic index, the lowest group had less than 70.8; the highest group had more than 77.7. As to the glycemic load, the lowest group had less than 147 units; the highest group had more than 234 units.

As compared to the lowest groups, the risks of ovarian cancer for the highest groups were 1.7 times higher based on the glycemic indices; 1.7 times higher based on the glycemic loads. The study suggested that an abnormally high circulation level of insulin, as a result of eating foods with high glycemic index and glycemic load, might have a role in developing ovarian cancer.

Glycemic Index And Breast Cancer

Silvera SAN, et al. "Dietary carbohydrates and breast cancer risk: a prospective study of the roles of overall glycemic index and glycemic load." International Journal of Cancer, Volume 114, Number 4, Pages 653-8. Apr 20, 2005. http://www.cababstractsplus.org/google/abstract.asp?AcNo=20053067002

This study observed a positive link between the risk of breast cancer and the glycemic index in the postmenopausal women. The fifth group of the postmenopausal women who took the highest glycemic index foods had 1.87 times higher risk of breast cancer than the first group who took the lowest glycemic foods.

Glycemic Load And Breast Cancer

Higginbotham S, et al. "Dietary Glycemic Load and Breast Cancer Risk in the Women's Health Study." Cancer Epidemiology Biomarkers & Prevention, Volume 13, Number 1, Pages 65-70, January 2004. http://cebp.aacrjournals.org/cgi/content/abstract/13/1/65

This study observed that women who had the highest daily glycemic load (consuming the highest amount of carbohydrates) and low levels of physical activities had a 2.35 times higher risk of breast cancer than the first group who had the lowest daily glycemic load. Also, the fifth group of premenopausal (before menopause) women who had the highest daily glycemic index of foods (consuming the foods with the highest glycemic index) and low levels of physical activity, had a 1.56 times higher risk of breast cancer than the first group who had the lowest daily glycemic index of foods.

Glycemic Load, And Colorectal Cancer Risk In Men

Michaud DS, et al. "Dietary Glycemic Load, Carbohydrate, Sugar, and Colorectal Cancer Risk in Men and Women." the Cancer Epidemiology Biomarkers & Prevention Volume 14, Number 1, Pages 138-147, January 2005. http://cebp.aacrjournals.org/cgi/reprint/14/1/138

For women, this study found no statistically significant relationship between the risk of colorectal cancer and the foods consumed, in terms of glycemic load, glycemic index, sucrose, and fructose. However, the study found a strong, positive link between the risk of colorectal cancer in men and the amount of daily glycemic load, as well as the consumption of fructose and sucrose. These risks were even stronger if men were overweight (with body mass index above 25 kilogram/m²).

Glycemic Load, And Women Colorectal Cancer

Higginbotham S, et al. "Dietary Glycemic Load and Risk of Colorectal Cancer in the Women's Health Study." Journal of National Cancer Institute, 2004; 96:229 –33. http://jnci. oxfordjournals.org/cgi/reprint/96/3/229

This study placed women into five groups based on glycemic load and glycemic index, as well as daily consumption of carbohydrates, non-fiber carbohydrates, and fructose. The study observed that, as compared to the lowest group, the highest group of daily glycemic load had 2.85 times higher risk of colorectal cancer. Also, the risks for the fifth group were 2.41 times higher for total daily consumption of carbohydrates, 2.60 higher for daily non-fiber carbohydrates, and 2.09 times higher for fructose, respectively. Although the result was not statistically strong, the risk of colorectal cancer for the fifth group was 1.71 times higher, in terms of daily glycemic index. The study pointed out the possibility of having higher risk of colorectal cancer for women who had a higher daily glycemic load.

Inflammation And Prostate Cancer

Nelson WG, et al. "The Role of Inflammation in The Pathogenesis of Prostate Cancer." the Journal of Urology Volume 172, Pages S6–S12, November 2004. http://www. ncbi.nlm.nih.gov/sites/entrez?cmd=Retrieve&db=PubMed& list_uids=15535435&dopt=AbstractPlus

We have considered several causes of cancer, including bacteria or viral infection, physical or chemical irritation, toxin, radiation, genes, foods, and et cetera. Speaking of foods, it does not matter which one of the causes mentioned above is found a tie to the development of cancer. More recent evidence has shown that all of them cause inflammation. As I have pointed out, hyperglycemia increases the level of inflammation inside the body.

This study reviewed articles about the relationship between inflammation and prostate cancer. The study found convincing evidence that inflammation played an important role in the development of prostate cancer. As inflammation of the prostate had a role in developing prostate cancer, the study suggested measures to prevent that inflammation might be helpful in prevention of the disease.

Carbohydrates And Prostate Cancer

Augustin LSA, et al. "GLYCEMIC INDEX, GLYCEMIC LOAD AND RISK OF PROSTATE CANCER." International Journal of Cancer. Volume 112, Pages 446–450. 2004. http://www.ncbi.nlm.nih.gov/sites/entrez?db=pubmed&uid=15382070&cmd=showdetailview&indexed=google

As more evidence has shown the association between inflammation and the risk of prostate cancer, we should suspect that dietary carbohydrates plays a major role in the development of prostate cancer. That is because the amount and grade of dietary carbohydrates influence the level of blood glucose, which affects the level of inflammation inside the body. This study found that the risk of prostate cancer had a direct link to dietary glycemic index and glycemic load.

On November 12, 2007, Dr. Stephen J. Freehand of Duke University Medical Center and his associates reported that a non-carbohydrate high-fat diet was able to delay the growth of prostate cancer and increase survival in a mice model study. He and his team planed to conduct a human clinical trial in the future.

Sugar And Lung Cancer

De Stefani E, et al. "Dietary sugar and lung cancer: a case-control study in Uruguay." Nutrition and Cancer, Volume 31, number 2, Pages 132-7. 1998. http://www. ncbi.nlm.nih.gov/sites/entrez?cmd=Retrieve&db=pub med&dopt=AbstractPlus&list_uids=9770725&query_ hl=2&itool=pubmed_DocSum.

This study showed individuals who consumed the highest daily amount of sucrose had 1.55 times higher risk of developing lung cancer than those who took the least daily amount. The risk would be higher in the cases of certain types of lung cancer than in others.

Hyperglycemia, Vitamin C And Cancer

Sargeant LA, et al. "Vitamin C and hyperglycemia in the European Prospective Investigation into Cancer--Norfolk (EPIC-Norfolk) study: a population-based study." Diabetes Care, Volume 23, Number 6, Pages 726-32. June 2000. http://care.diabetesjournals.org/cgi/reprint/23/6/726.pdf

Hyperglycemia plays an important role in inflammation and development of cancer.

This study found the level of Vitamin C was higher in the participants who had normal blood glucose than those who had hyperglycemia or diabetes (DM). Therefore, the higher the blood glucose was, the lower the level of the Vitamin C would be.

Glycemic Control, Vitamin Supplement, And Cancer

Krone CA and Ely JTA. "Controlling Hyperglycemia as an Adjunct to Cancer Therapy." Integrated Cancer Therapies, Volume 4, Number 1, Pages 25-31. 2005. http://ict.sagepub. com/cgi/content/abstract/4/1/25

This article reviewed the level of blood glucose in relation to the status of cancer. The study found that patients who were in active stage of cancer had a higher level of blood glucose and lower level of Vitamin C than healthy individuals. On the other hand, the cancer patients who were in remission had lower level of blood glucose and higher level of Vitamin C. The article suggested that cancer therapy should include controlling blood glucose and supplying Vitamin C to patients.

Hyperglycemia And Cancer

Ely JTA. "Glycemic Modulation of Tumor Tolerance." Journal of Orthomolecular Medicine, Volume 11, Number 1, Pages 23-34. 1996. http://faculty.washington.edu/ely/JOM1. html

This article pointed out that the immune system, which constantly removes cancer formation in the body, is suppressed as the level of blood glucose rises. The article suggested that controlling blood glucose level should be part of cancer therapy.

CARBOHYDRATES AND LONGEVITY

In world history, there are stories about kings and emperors who wanted to live longer and forever. They commanded their wise men to search for the ways to extend their life expectancy.

To no surprise, they failed to live longer and forever. In many cases, they actually shortened their lives by eating different specially cooked dishes, which were delicious, and very likely sweet too. But why did they fail to have the life they wanted, in spite of their unlimited power in making everything else possible? From studies about the impact of carbohydrates on the life of living cells, we begin to realize that unlimited carbohydrate consumption or hyperglycemia can damage the cells and promote their early death!

Calorie (Carbohydrate!) Restriction For Health

Fontana L, Meyer TE, Klein S and Holloszy JO. "Long-term calorie restriction is highly effective in reducing the risk for atherosclerosis in humans", <u>Proceeding of the Academy of Science of the United Sates of America</u>, Volume 101, Number 17), Pages 6659–6663. April 27, 2004. http://www.pnas.org/cgi/reprint/0308291101v1.pdf?ijkey=8217c95da53 9e98b528b496ad717435148ab5ea7

This study reported that cutting down the daily total number of calories from the typical American diet helped improve the blood test results that are the indicators for excellent health. While the reduction of the total calories was remarkable, the reduction of the daily amount of carbohydrates was even more remarkable. The improvement was observed among individuals at all ages. This implies that it is never too late to switch your dietary style for an improved, lasting health for the rest of your life!

High Blood Sugar And High Death Rates

Shaw JE, et al. "Isolated post-challenge hyperglycaemia confirmed as a risk factor for mortality," <u>Diabetologia</u>,

Volume 42, Number 9, Pages 1050-4. September 1999.
http://www.ncbi.nlm.nih.gov/sites/entrez?cmd=Retrieve&db
=PubMed&list_uids=10447514&dopt=AbstractPlus

This study observed that individuals with high post-challenge blood sugar (200 mg% and over), in comparison with those without high post-challenge blood sugar, had a higher death rate for all causes; 2.7 times higher for men and 2.0 times higher for women. They also had a higher death rate for cardiovascular diseases; 2.3 times higher for men and 2.6 times higher for women. Incidentally, the men with high post-challenge blood sugar also had 8 times of death rate for cancers as compared to those men who had normal post-challenge blood sugar.

Chapter 7: In My Opinion

I went to pharmacy school in September 1961 and Nagasaki University School of Medicine in April 1967. I have been in the medical field for 47 years and I have come to understand more about the failure of today's medicine. It fails to teach us how to keep up our health, to prevent us from contracting diseases in the first place. As a result, we have become a people who spend too much of our financial resource on medications and surgeries, for a quick fix.

A quick fix for our medical complaints does not address the source of our illness for removing the cause(s) of the illness. Rather, a "quick fix," as it is termed, is just a short symptomatic relief, like painting over a cracked wall without correcting the cause(s) for the crack. Eventually, the crack will reappear, and even grow bigger, and it may be not repairable next time. For example, a patient has acid reflux, so he asks his physician for help. After a physical examination, he will likely receive antacid, either by prescription or in an over-the-counter medication. When we realize that cases of acid reflux are on the rise over the past decades, we should ask what makes the increase. Is there any other health problem on the rise too? Sure, there are many related health problems on the rise. To name a few, these are obesity, eating too many sweets, including starchy foods and sugar, and et cetera. Worst of all, recent studies have found a link between cancers of the esophagus and excessive eating of carbohydrates!

During the same decades, we have had many more new cases of diabetes (DM) and cardiovascular diseases; the latter includes high blood pressure, heart attacks, and strokes. We have seen more cases of arthritis, allergies such as the peanut allergy in children, which was rare during my childhood. We have also had more neurological and mental disorders. These disorders include Parkinson's disease (PD), multiple sclerosis (MS), Alzheimer's disease (AD) and dementia, seizures, attention deficit disorder (ADD), autism, and et cetera.

Also during the same decades, we have advanced medical technology. We have gained more information about diseases. We can argue that we discover many new diseases solely because of such achievement. But why can't we prevent the diseases or at least reduce the number of new cases? That is because we have failed to recognize the real culprit, offender, or cause for the diseases and to find a sensible way for prevention. For your reference of my writing below, please use the number(s) in the parentheses for the article(s) of the reading list that I organized and posted online at www.carbohydratescankill.com.

"Carbohydrates Can Kill"

Since the middle of the 1900s, physicians and nutritionists have obviously misunderstood the role of three macronutrients affecting our health. These macronutrients are carbohydrates, fats, and proteins. They thought that dietary fats were the main source of our body fat. They thought carbohydrates should be the necessary source for our body energy. Of course protein was undoubtedly the source for re-supplying our body protein in the muscles, organs, and tissues. Their worst mistake came in encouraging people to eat carbohydrates, without a slightest suspicion of its terrible impacts on our health. They thought that carbohydrates were harmless. They are impressed by the low amount of calories, 4 kcal per gram, of carbohydrates. For that, they have strongly recommended carbohydrates as the main energy source for the body. They even paint fats as the most undesirable source of nutrition for us, in comparing fat with carbohydrate and protein. Hundreds of medical research articles, in both Chapter 6 and on the reading list posted online for this book, have told us a different story. We should realize that we have been wrong about the roles of carbohydrates in our metabolism and disease development.

As in the cartoon below, most diseases, if not all, are directly or indirectly caused by too much blood sugar. Blood sugar comes from glucose, fructose, and other sources. Most diseases are the branches of a wild tree growing out of the ground, or our body. The name of the wild tree is "High Blood Sugar After Meal."

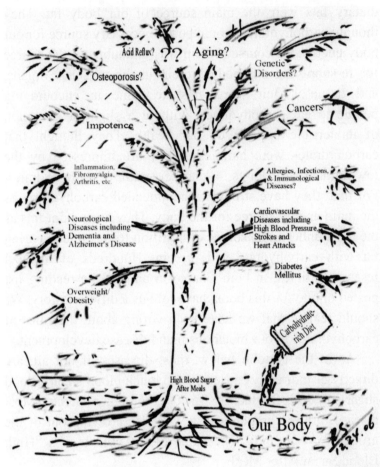

Figure 36. Diseases Caused By High Blood Sugar After Meal

On the one hand, we want to cut the branches and the trunk down with medications and surgeries. That is because we have trouble(s) with our health when illnesses grow quickly out of control. On the other hand, we keep feeding the ground, or, our body, with plenty of "plant feed", which is a carbohydrate-rich diet. Undoubtedly, the "plant feed" makes the wild tree grow faster than we can cut its branches and trunk down to keep them under control. The more carbohydrates (excluding fiber) we eat, the higher the level of our blood sugar will be after eating. The best way to control the growth of the wild tree is to limit the "plant feed", which is rich in carbohydrates, to the ground, or our body. With limited "plant feed" to the ground, the wild tree cannot grow or, at least not so fast! With the illustration, we can understand that excessive blood sugar (glucose) can cause different problems. These problems do not need to happen at the same time in all cases. Obesity is a good example. It happens when one has eaten too much food rich in carbohydrates and likely far more calories than he needs. He may have or may not have other problems such as diabetes (DM), arthritis, or other diseases. All of a sudden, he finds out that he has a cancer. At that time, he may also have diabetes (DM), and probably some other diseases, too.

The case in point is metabolic syndrome, a name physicians would give to a combination of problems with advanced diabetes (DM), insulin resistance, and cardiovascular diseases. As said, metabolic syndrome was originally called syndrome X, metabolic syndrome X or insulin resistance syndrome. From the meaning of metabolic syndrome, we seem to blame the failure of the metabolic process of our body for the syndrome. We thought that our body was supposed to take in whatever amount of food it needed, especially carbohydrates, and that it should be able to handle the foods well. When we found that we had problems, we blame our pancreas for the failure. We suspected that our pancreas must have some sorts of genetic disorders, immune defects, infections by bacteria and/or viruses, and toxic chemicals or biological toxins. Why can't we prevent ourselves from contracting these diseases? Because most medical doctors and nutritionists are unwilling to admit that eating excessive carbohydrates is harmful. Because they do not think that carbohydrates are the targets, there is no prevention for us from having the diseases. Eating excessive carbohydrates has harmed our health and drained our financial resources. We all must realize that CARBOHYDRATES CAN KILL! Then, we can educate ourselves to properly realign the portion of each macronutrient for good health and disease prevention.

Speaking Of Diabetic Diet

Until the recent years, physicians had always treated diabetes (DM) first with a restricted diet. The diabetic diet cuts the amount of both calories and fats. It is still rich in carbohydrates. Most physicians believe the "advantage of low calories for carbohydrates in comparison to fats, gram for gram." They even encourage the diabetic patient to eat more carbohydrates in place of fats. They thought that the patient would be able to eat more in volume and less in calories. They thought that the total amount of calories was the cause of obesity, thus, diabetes mellitus. They should understand that a diabetic patient could be thin during his entire life. They did not realize that the diabetic diet was probably worsening the patient's diabetes (DM) in the first place.

There is no organ or tissue of our body that can be overworked or take repeated assaults without a consequence. This causes injury to the organ and tissue. As a result, the injured organ and tissue will fail and die. The pancreas is no exception. Repeated rises in blood glucose assaults the beta-cells of the pancreas. Sooner or later, the pancreatic beta-cells have been so damaged that they die. At that time, we find out that we are diabetic.

The patient's pancreas cannot handle the continuing increase of blood glucose from the carbohydrate-rich diabetic diet. This high amount of blood glucose demands more production of insulin by the pancreas. The pancreas will be exhausted sooner or later. Studies have shown that a high amount of blood glucose can damage the body, organs, and tissues, including the pancreas. Inflammation, destruction, mutation of the cells, glycation, and other mechanisms cause the damages. There should be no doubt that a diabetic diet, with no restriction on carbohydrates, is not only ineffective, but actually harmful. (See CARBOHYDRATES AND DIABETES MELLITUS, Chapter 6.)

One of our terrible misconceptions is that we cannot live without carbohydrates. Anssi H. Manninen found that the body does not require a minimal amount of glucose or insulin. Our brain, heart, and skeletal muscles can use beta-hydroxybutyrate well. Beta-hydroxybutyrate is a ketone body from breaking fats down, when the body uses a very low carbohydrate diet. Besides, many cells of our body do not need insulin to pick up glucose from the blood circulation. These cells are red blood cells, nerve cells (neurons), liver cells, some tissues of the bowel, and the beta-cells of the pancreas. (Manninen AH. "METABOLIC EFFECTS OF THE VERY-LOW-CARBOHYDRATE DIETS: MISUNDERSTOOD "VILLAINS" OF HUMAN METABOLISM." Journal of the International Society of Sports Nutrition.. Volume 1, Number 2, Pages: 7-11. 2004.)

Thus, our body does not need so much carbohydrate. This contradicts the recommendations in the official food pyramid for maintaining our daily activities and health. Eating too many carbohydrates causes illnesses and speeds up the aging process.

A patient with insulin-dependent, the unrestricted carbohydrate diet can worsen his diabetic symptoms and complications. This is true even when the patient follows the instructions for taking insulin injections.

Physicians use abnormal fasting blood glucose level and/or two-hour blood glucose level after a glucose tolerance test for a diagnosis of diabetes mellitus. The blood glucose tolerance test involves having the patient drink a mix of 1.75 grams of glucose per kilogram of his body weight, up to 75 grams of glucose. The blood glucose level at the two-hour mark is used for diagnosis. But either or both tests do not tell the whole story how the body fares. Instead, we should monitor the blood glucose level at all times. In my opinion, a reasonable blood glucose level should be 150 mg% or lower after a meal. This 150 mg% represents 150 mg of glucose in every 100 ml of blood or serum. The level of blood glucose beyond 150 mg% could damage the body, organs, and tissues. The possible result of such damages includes most, if not all, of our popular and rare diseases.

When a patient is found to be diabetic or pre-diabetic, the first thing he should do is restrict the amount of carbohydrates in his diet. He must reduce the numbers of both the glycemic index and glycemic load of food he consumes.

As in the cartoon below, the house (the patient's body) is on fire, a blaze that diabetes mellitus and its complications have caused. The hydrant is pumping diabetic medications into the house (or patient) to fight the diabetic fire. The medications include insulin, in the case of insulin-dependent DM. However, there is no carbohydrate restriction in the order and the patient continues carrying more fire logs into the house for the blaze. The fire logs here are rich in carbohydrates. We should expect that the house is going to burn down soon.

Figure 37. Speaking of Diabetic Diets

Gastric bypass is now a popular procedure for helping to stop the trend in obesity. In the meantime, reports showed unexpected improvement in the patients' diabetes (DM) after gastric bypass. The improvement began a few days after the gastric bypass surgery or before the beginning of weight loss. The reports should make us reconsider the link between obesity and diabetes mellitus. Does obesity really cause diabetes mellitus? Rather, as I pointed out in the first topic of this chapter, both obesity and diabetes mellitus are results of a high blood glucose after meal. An obese person is not necessarily diabetic for a while. On the other hand, a diabetic patient is not necessarily overweight or obese at any point of his life; he could be underweight for his entire life.

Although gastric bypass surgery has shown a drastic improvement in the blood glucose level, there are risks in that surgery. Carbohydrates in our foods include polysaccharide, disaccharide, and monosaccharide. All polysaccharide and disaccharide must be converted into monosaccharide before it can be absorbed by the small intestine. Monosaccharide is the source of blood glucose. The duodenum receives pancreatic enzymes, which convert polysaccharide and disaccharide into monosaccharide. Gastric bypass lets the carbohydrate foods bypass the duodenum and cuts short the time needed for the pancreatic enzymes to meet the polysaccharides and disaccharides. Most polysaccharides and disaccharides are moved into the large intestine without being converted into monosaccharide. As a result, a smaller amount of monosaccharide enters the blood circulation, the blood glucose level is lowered, and the pancreas gets time off for recovery. This is just a case of forced dieting with carbohydrate restriction by surgery. More reports should come out in later years that gastric bypass also improves the risks for coronary heart disease, stroke, arthritis, dementia and Alzheimer's disease, Parkinson's disease, cancers, and many more. Most of the diseases are a result of hyperglycemia from unlimited carbohydrate consumption. Widespread research has shown improvement in a patient's diabetic condition with a restricted carbohydrate diet. The participants in these tests needed fewer or no medications for diabetes (DM) at the end of the experiments. Studies found that hyperglycemia damages the tissues and organs, including the pancreatic ß (beta) islet-cells. Also, hyperglycemia mutates cells, including the pancreatic cells. With all the studies, we should question the wisdom of claims that diabetes mellitus is a genetic disease. Because of the false claim, physicians focus on treating the diabetic patient for his symptoms rather than seeking the cause of his disease.

In the issue of Diabetes Care, Volume 31, Pages S5-S11, January 2008, the American Diabetes Association issued its Executive Summary: Standards of Medical Care in Diabetes—2008. The summary includes the use of a low-carbohydrate diet under the section for Medical Nutrition Therapy, without its endorsement. Its new position statement is still disappointing. However, it has made a small but important change in setting up recommendations about how we may prevent and manage diabetes (DM).

Inflammation, Inflammation & Inflammation!

Over the past decades, we have considered inflammation to be a necessary and normal function of the body. Inflammation happens in response to different situations, including radiation, tissue injury, invasion by foreign organisms such as bacteria and viruses, or irritation by foreign particles or substance. Inflammation is supposed to promote healing or repairing of the injured and damaged tissue. In the meantime, we have also realized that inflammation plays an important role in several diseases such as arthritis and fibromyalgia (or, fibromyositis). Some recent studies have also found the link between inflammation and other disorders, such as psoriasis and osteoporosis.

During recent years, we have increasingly understood both acute and chronic inflammation. Inflammation is the major cause of cardiovascular diseases, including arteriosclerosis (hardening of the arteries), atherosclerosis (damage of the blood vessels as a result of arteriosclerosis), heart attack, and stroke. Inflammation also causes many disorders that may involve the gastrointestinal tract including the pancreas and liver, the urinary system including the kidneys and bladder, and the genital (sexual) organs. They may affect the respiratory system including the sinuses, larynx, trachea and bronchus, the lungs, and the endocrine system (the system of hormones). More studies have also found that inflammation is a possible suspect for many, if not all, kinds of cancers and degenerative disorders of our nervous systems. Therefore medical researchers should work harder in the direction of clarifying the exact role of inflammation in the process of developing diseases. After all, inflammation may not be necessary in the healing process of the body.

When the body is injured, its tissue is damaged, whether we have an open or closed wound. We notice swelling, skin redness, higher temperature or fever, and pain or soreness at the site of injury. These four symptoms are the typical picture of an acute inflammation. They present in response to tissue injury, in case of either an open or closed wound.

As time passes, in the case of a closed wound, we observe that the swelling goes down, the skin redness fades away, the temperature returns to normal, and the pain decreases then subsides. Now we know that acute inflammation has gone away, after completing the process of tissue healing.

On the other hand, in a case of an open wound without infection by bacteria or viruses or contamination by foreign substances, we probably observe a similar course of recovery. The only difference is that we also observe the process of closing up the open wound.

In case of infection or contamination in the course of recovery from an open wound, we see complications. Infection involves bacteria, viruses, and fungi, and et cetera. The complications include delayed healing, lots of discharge and pus, more swelling, fevers of the wound (local fever) and the body (systemic fever), and an extended period of feeling pain. Also, we see poor scarring and keloid formation or keratosis.

It is a fact that bacteria, viruses, and foreign bodies are present everywhere, but why do some of us heal faster and better than others do, in a similar situation of either a closed or open wound? Apparently, every body has a distinct level of defense against infection and contamination that is different from all others. So do the levels of inflammation from one person to another. The person who has a higher level of inflammation responds much more seriously to tissue damage. As said, tissue damage is a result of infection or contamination.

For many diabetic patients, a wound will not heal

quickly and easily, even with the newest treatment and the most powerful antibiotics. This can be a nightmare for the treating physician. Studies have confirmed that the body of diabetic patients has a higher level of inflammation. Without lowering the diabetic patient's level of inflammation, the wound is very hard to heal. Then, we should ask why the diabetic patients have a higher level of inflammation. We should blame the common problem among diabetic patients. That is the abnormally high blood glucose level! Keeping the blood glucose level of the diabetic patient within the normal range at all times will reduce the level of inflammation; it is the utmost important factor for successfully healing wounds.

The above examples are wounds of our skin and flesh, which are within our own ability to take care of and monitor. However, similar situations occur inside the body and its organs, too. Often, we do not even realize that an injury or damage to our organ(s) has taken place. Eventually the damage is serious enough to develop certain symptoms. By that time, we might or might not be able to help heal the wound of the damaged organ(s).

Some of our white blood cells fight bacteria, viruses, and foreign bodies by swallowing them up. This is phagocytosis. The strength of phagocytosis by our white blood cells is inversely tied to the level of blood glucose; the strength is highest when the level of our blood glucose falls between 50 gm% and 100 mg%; it is significantly weakened and lowered as our blood glucose level goes up. (365) This is why we observe lots of pus in the wound of a diabetic patient. Pus is the debris of our white blood cells. The white blood cells did not have much of a chance to fight off infection in the first place, when the patient had such a high blood glucose level. The total count of our white blood cells also increases as the level of our blood glucose goes up from the normal range of blood glucose. (497) An increasing white blood cell count may be a response to a chronic inflammatory condition from a high blood glucose level. Alternatively, the increase may be an acute response, an attempt to increase the number, rather than the quality of white blood cells. This is to improve their strength in defending the body from tissue damage, infection, and injuries.

There are positive links among the level of inflammation, the blood glucose level, and the white blood cell count. These links make me ask if leukemia (cancer of the white blood cells) is a result of inflammation. A study report by Christine M. Kausum and her research team supports my suspicion. The study reported that the postmenopausal women who took aspirin regularly, had much lower risks for some subtypes of leukemia than those who did not. However, the same study also found that non-steroid anti-inflammatory drugs did not decrease the risks the way aspirin did. (384)

The bone marrow produces blood cells. I suspect that if acute or chronic inflammation happens and damages the bone marrow, the marrow would be unable to produce blood cells, as it should. Is not it possible that leukocytopenia (shortage of white blood cells), thrombocytopenia (shortage of platelets which is an important cell for blood clotting), aplastic anemia (shortage of red blood cells), and other diseases are a result of damaged bone marrow by inflammation?

Studies have also found that chronic inflammation is

the main suspect of causing cancers. It is true that bacteria, viruses, chemical substances, physical irritation, radiation, and others might be linked to the development of cancers. However, all of them trigger or cause inflammation. To support our suspicion, studies observed aspirin and other anti-inflammatory drugs lowered the risks of cancers in many different parts of the body such as the pancreas, colon, stomach, prostate, lungs, and many other organs. (394, 395, 396, 397, 398, 400, 402, 926, 931, 953, 954, 955, 956, 959, 975, 976, 981, 982, 991, 992, 993)

Acute inflammation "may be" a needed response in the healing of tissue injury and damage. It is just "may be," in my opinion. Excessive inflammation and chronic inflammation are harmful to our health. Excessive and chronic inflammation cause more injury to our tissues. They can cause mutation of normal cells to become malignant or cancer cells. Moreover, they probably help spread or metastasize cancer. (391, 912, 941)

In the meantime, studies have also found a positive link between hyperglycemia (high blood glucose) and the risks of cancers. (893, 894, 895, 896, 897, 898, 899, 900, 901, 902, 903, 904, 905, 906, 1042) Most of the blood glucose is from carbohydrates, excluding non-absorbable fibers, in our diet. Of course that was no surprise to me. The studies have also linked eating an unlimited amount of carbohydrates to the risk of cancers.

Most importantly, we must know high blood glucose causes inflammation, in both the acute and chronic situations. (365, 366, 375, 381, 382, 383) To prevent inflammation in the body, one must prevent the level of blood glucose from rising to a dangerous point. To maintain a safe level of blood glucose, one must avoid eating an unlimited amount of carbohydrates. Is eating lots of fruits a good thing for our health? Not really!

In history, emperors, kings, queens, and rich people tried to find ways to live longer, or even forever. Although we are sure that no one can live forever, we can live longer with a healthy lifestyle. The body's normal cells have a regular life span and a scheduled death (apoptosis). Studies show that inflammation or hyperglycemia can affect the behavior of our cells in the process of apoptosis. (387, 570, 1161) Inflammation or hyperglycemia can damage our normal cells and make them die sooner. When the cells do not die the way nature intended, they become cancer cells and reproduce, and spread (metastasis).

Glycation is a simple chemical process that carbohydrate or sugar (including fructose, galactose, and glucose) bonds itself with either fat, protein, or both. The product is glycation product. The primary glycation product can further bond with more protein or fat to become advanced glycation endproducts (AGEs). AGEs are toxic to our tissues and cause inflammation inside our body.

Glycation can happen in either inside or outside of our body. When AGEs are produced in cooking food, 30% of them can be absorbed through digestion. On the other hand, fructose, galactose, and glucose inside our body can initiate glycation as well. Several degenerative diseases especially such as Alzheimer's disease, Parkinson's disease, amyotrophic lateral sclerosis (ALS or Lou Gehrig's Disease), and multiple sclerosis (465, 467, 468, 763, 764, 776), and arthritis. (619, 815) Interestingly studies have also found a positive link between hyperglycemia (high blood sugar) brought on by a carbohydrate-rich diet and the severity of arthritis. (115, 618)

So many ideas illustrate the tie between inflammation and longevity. Lowering the level of inflammation of our body at all times is the most important and effective way for a healthy and long life.

An Unfair Race

Now we know that lowering the level of inflammation of the body is the top priority for keeping us healthy. Infomercials have told us that oxidation of our cells causes injuries and damages to our cells, tissues, organs, and body. Oxidation is a chemical reaction. Oxidation causes aging of the cells, thus, our life. In oxidation, electrons are transferred from a substrate to an oxidizing agent. Antioxidants are supposed to accept the electrons or the free radicals to become oxidized. Antioxidants are supposed to protect the components of our cells from oxidation.

Some of us may not really understand the meaning of oxidation. However, we have been told that we should take antioxidants such as vitamin C, vitamin E, glutathione, superoxide dismutase, catalase, and peroxidase. Studies also show that taking some anti-inflammatory medications can reduce the risks for different diseases including cardiovascular diseases, inflammatory diseases, pancreatitis, Alzheimer's disease, and even cancers. Then we shall live a healthy and long life. Not so fast!

First, we must know why our body continues to have inflammation and oxidation. Without knowing the reason(s), we cannot be sure that the daily amount of antioxidants and anti-inflammatory medication we take would be enough to protect us from developing diseases or shortening our life. Studies have found that oxidation inside our cells is linked to inflammation. (642, 816, 820) Research also shows a positive link between hyperglycemia and inflammation, in both acute and chronic situations.

We can only take so many antioxidants and anti-inflammatory medications each day. At the same time, we continue to feed ourselves with an unlimited amount of carbohydrates. That increases the blood glucose level and the level of inflammation. Eating carbohydrates with high glycemic indices is even worse. We should expect that the level of inflammation inside our body would rise as our blood sugar level goes up. How can we expect a daily dosage of antioxidants and anti-inflammatory medications to protect us from developing diseases or shortening our life?

This is an unfair competition on the racetrack inside the body. On the one hand, we allow the first racer a fixed amount of fuel in the race. On the other hand, we continue to refuel the other racer without a limit. Antioxidants and inflammatory medications would fail to make us healthier in such an unfair race!

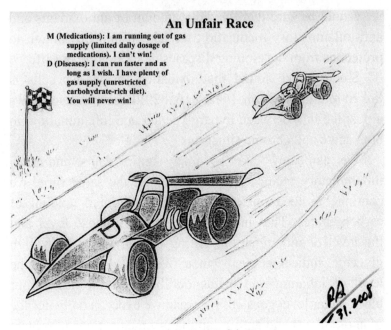

Figure 38. An Unfair Race

We cannot ignore the side effects of antioxidants and anti-inflammatory medications. Studies found favorable effects on cancer prevention with anti-inflammatory medications. But we should not try the long-term use of anti-inflammatory medications for cancer prevention, because of their side effects.

Studies found that antioxidants could become harmful to our body. Antioxidants receive free radicals (oxidized products) at the same time, and they eagerly pass on the free radicals to others, including the substance inside our cells. This is just like a volleyball game. In this situation, the substance in our cells end up oxidized and damaged after all. (1012, 1135) For example, vitamin C is supposed to be used as antioxidant. But, vitamin C, oxidized products, and advanced glycation endproducts can cause unfavorable effects that include initiating glycation. (874) Glycation, mentioned elsewhere in this book, can damage our tissues and organs to cause arteriosclerosis, diabetes mellitus, Alzheimer's disease, cancers, and many other diseases. Without preventing oxidation inside our body, taking antioxidants is only briefly helpful at best. Sometimes, it could be harmful instead.

Limiting the amount of carbohydrates in our everyday diet would decrease the blood glucose; lowering the blood glucose would reduce the level of inflammation and oxidation. In turn, we can live a healthy and longer life.

The Role Of Hyperglycemia

Most of our blood sugars come from carbohydrates, excluding the non-digestible fibers, in our foods. The more carbohydrates we eat, the higher the level of our blood sugar will be. Most of our blood sugar is glucose. Blood glucose is very easily used in producing energy for the body. However, more glucose (hyperglycemia) than our body needs for growth (in case of young children and adolescents) and daily activities can hurt the body in many ways. Here are just a few describers of potential problems caused by hyperglycemia: (1) inflammatory and proinflammatory; (2) prothrombotic; (3) vasoconstrictive; (4) arteriosclerotic and atherosclerotic; (5) glycosylative and pro-glycation.

Glucose and other sugars from carbohydrates cause inflammation inside our cells, tissues, organs, and body. Chronic and acute hyperglycemia, and even just a case of hyperglycemia for a brief period after eating more carbohydrates, can increase a host of inflammatory factors.

These inflammatory factors include C-reactive protein and inflammatory cytokines. Because these inflammatory factors usually are generated and released in response to tissue damage, in my understanding, glucose likely causes inflammation. In addition to existing inflammation, glucose can add more inflammation, therefore, is proinflammatory. (384)

The process of healing from a wound is very difficult for a diabetic patient when his blood glucose is uncontrollably high. A non-diabetic patient would also have a slower healing process from irritation, inflammation, or infection, when he has eaten too many sweets. In my experience, staying away from sweets could help prevent or recover from a mild episode of infection or irritation.

Hyperglycemia can trigger biochemical chain reactions in blood coagulation or blood clot formation. Therefore, hyperglycemia is prothrombosis. When a blood clot forms over an open wound, it can help prevent blood loss and stabilize the vital signs (thrombosis.) If blood clot formation happens inside a blood vessel, it can partially or totally close up the opening of the blood vessel, thus reducing or stopping the blood flow to the tissue(s) and organ(s). Sometimes, it causes death suddenly or later. Cerebral infarction and myocardial infarction are typical accidents that result from blood clot formation. Cerebral infarction is a type of strokes due to lack of circulation to the brain. Myocardial infarction is a type of heart attacks as a result of lack of circulation to the heart muscles.

The inner layer of the artery (endothelium) produces and releases nitric oxide. The chemical symbol of nitric oxide is "NO." Nitric oxide relaxes the muscle layer of the blood vessel, and opens up the blood vessel (vasodilatation). Consequently, vasodilatation eases the blood flow and it reduces blood pressure. Hyperglycemia decreases the production and release of nitric oxide from the endothelium. With the smaller amount of nitric oxide for relaxing the muscle layer of the blood vessel, the opening of the blood vessel decreases (vasoconstriction.) Vasoconstriction reduces the caliber of the blood vessel and so reduces blood flow, then the blood pressure goes up (hypertension). This shows why a diabetic patient ends up with hypertension. Many of us have both hyperglycemia and hypertension. But, we pass the screening for diabetes (DM) because of the flawed criteria for making a diagnosis of diabetes (DM). By the time we are diagnosed with diabetes (DM), the disease has advanced much farther. Early detection of diabetes (DM) can save us from having a series of serious health problems.

A case of particular importance to both men and women is erectile dysfunction (ED) or impotence. Often, men blame aging for their difficulty in keeping an erection during intercourse. Likewise, women blame aging for their loss of sexual excitement. They use medications for the sake of a 36-48-hour relief from the symptoms. They do not know that erectile dysfunction is possible a result of hyperglycemia and vasoconstriction. A majority of them with erectile dysfunction may have more serious cardiovascular diseases. They should not opt to using medications for temporary relief of erectile dysfunction. Rather, they should consult their physicians for earlier cardiovascular disease screening. The early consultation can save them from shortening their lives.

The publicity on the link between cholesterol (especially

low-density lipoprotein cholesterol or LDLc) and arteriosclerosis (and atherosclerosis) gains our attention. Most of us know that we should watch out the amount of both cholesterol and LDLs. However, VLDLc (very low-density lipoprotein cholesterol), a member of the LDLs, is the worst lipoprotein cholesterol and is responsible for the development of arteriosclerosis.

Diabetic patients have a high risk of developing arteriosclerosis. Hyperglycemia is responsible for producing VLDLc, in addition to triglycerides. With hyperglycemia, inflammation happens on the inner layer of the blood vessel (endothelium), which attracts LDL cholesterol, especially VLDL cholesterol, to deposit cholesterol to the endothelium. As the deposit of cholesterol continues in hyperglycemia, the wall of the blood vessel becomes hardened or rigid (arteriosclerosis.)

A normal blood vessel has very small blood vessels running inside its wall for supplying oxygen and nutrition. While the deposit of cholesterol in the endothelium (arteriosclerosis) becomes a thick plaque, the wall of the blood vessel wants to extend its blood supply to the plaque (revasculization.) Unfortunately, the cholesterol plaque is so hard and does not allow revasculization, and damages the endothelium (atherosclerosis.) Sometimes the cholesterol plaque from arteriosclerosis or atherosclerosis can separate itself from the blood vessel wall and block the opening of the blood vessel(s). When that happens, stroke, heart attack, and other types of thrombosis follow.

In some cases, atherosclerosis causes so much damage to the wall of the blood vessel that the wall of the blood vessel becomes very thin and easily bulged by the pressure of the blood flow. The bulged blood vessel wall is called aneurysm, which can be ruptured and cause acute blood loss and possible death.

Keep this in mind. Hyperglycemia happens not only in diabetic patients. But hyperglycemia also happens in someone like many of us, who "have never been diagnosed" with diabetes (DM). A brief episode of hyperglycemia, after eating too much of a "sweet goodie,' could cause an acute episode of inflammation inside the blood vessel, with vasoconstriction and hypertension. Suddenly, stroke or heart attack follows. This is the reason why we cannot completely exclude people who do not have history of diabetes (DM) from the risk of stroke, heart attack, or other diseases.

Glycation is a simple bonding reaction between sugar and fat or protein to produce glycolipid or glycoprotein. Glycolipid is a molecule of carbohydrate and lipid or fat. Glycoprotein is a molecule of carbohydrate and protein. The bonded molecule, either glycolipid or glycoprotein, is called a glycation product. A glycation product can initiate further glycation with either fat or protein to form large molecules or advanced glycation endproducts (AGEs).

Glycation changes the nature of the original fat and protein molecules, subsequently damaging the tissues and organs and speeding up the aging process. Studies have already implicated glycation in the development of diabetes (DM), cardiovascular disease such as arteriosclerosis, cancers, degenerative neurological disorders such as Alzheimer's disease, ALS, multiple sclerosis, and others. (875) Studies have also found a patient with diabetes mellitus (or hyperglycemia) has a high risk of developing Alzheimer's disease and other degenerative neurological disorders. (769, 867, 868, 869)

For instance, recent studies have found insoluble plaque of amyloid-ß (abeta or Aß) in the brain of a patient with Alzheimer's disease. From amyloid precursor protein, the brain produces amyloid, which is soluble in water. In the meantime, the water-soluble amyloid is broken down for removal by protease, an enzyme to break down proteins. Studies showed that hyperglycemia in the brain can increase the amount of amyloid precursor protein (APP). (884, 1163)

Based on my review of studies, hyperglycemia increases the amount of amyloid precursor protein, and thus increases the amount of soluble amyloid. In the presence of hyperglycemia and glycation products or advanced glycation endproducts (AGEs), amyloid molecules bond themselves to form larger molecules, which are insoluble in water and cannot be broken by protease for removal. These are the amyloid-ß plaques found around the brain nerve cells (neurons). The plaques damage the neurons and develop Alzheimer's disease. (872, 873, 875, 876) This may explain why individuals with diabetes (DM) and undiagnosed hyperglycemia have higher risks of Alzheimer's disease and other neurodegenerative diseases.

Generally, the level of glycation is tied to the amount of sugars, especially blood glucose, available for the bonding reaction. For instance, the reading of hemoglobin A_{1c} reflects the level of averaged blood glucose for the last 4-6 weeks. The higher the blood glucose level, the higher the reading of hemoglobin A_{1c}. However, a stable and lower hemoglobin A_{1c}, does not remove the risk of an acute episode of stroke or heart attack, because an acute and sharp increase of blood glucose level, after eating excessive carbohydrates, can trigger an acute inflammation and blood clot formation inside the blood vessel. As mentioned above, hyperglycemia has a role in at least five pathological processes. Conventional medicine will likely suggest that we should take (1) anti-inflammatory drugs such as aspirin, (2) anti-thrombotic such as blood thinners or anti-coagulants, (3) vasodilatives such as medication for open up blood vessels, (4) anti-cholesterols or cholesterol-lowering drugs, and (5) new medications for promoting the breakdown of individual protein and fat for anti-glycation. While I do not dispute the logic behind the knee-jerk-reaction type of practice, I believe we should have alternatives.

Studies have shown positive results with ketogenic diets in experiments with treating cancers, Alzheimer's disease, uncontrollable epilepsy in children, and others. The ketogenic diet is a diet that contains little or no carbohydrates and a higher percentage of fats. Without or with little carbohydrates, the ketogenic diet produces ß-hydroxybutyrate, which is a kind of ketone body. Different from acetoacetate, the ketone body in the case of diabetic acidosis, ß-hydroxybutyrate can be used by the brain, heart, and skeletal muscles as the source of energy.

Also, evidence shows that hyperglycemia, a result of eating too many carbohydrates, can either cause or aggravate diseases such as osteoporosis, polycystic ovarian syndrome, chronic obstructive pulmonary disease (COPD), and many others. Learning the role of hyperglycemia should prompt us to lower the level of our blood glucose in the first place, for reducing or preventing the risks of diseases. Carbohydrates in the diet are the major source of blood glucose. Cutting the amount of carbohydrates in our diet definitely lowers the level of blood glucose. Consequently, the carbohydrate-restricted diet should reduce the risks of developing diseases.

Body Mass Index vs. Waistline

The average body weight of the population in both the United States and the world is increasing sharply. More and more wake-up calls are asking people to watch their weight and stop the trend of gaining.

Most commonly, we use Body Mass Index or BMI to decide if we are underweight, normal weight, overweight, and obese. BMI is a product by dividing our body weight over the square of our height in the metric system. (Please see Figure 19 Body Mass Index.) The index is same if we use a metric or the customary English system. There is no reasonable basis to explain the result of BMI for each individual. The index just reflects the relationship between the weight and the height for each person, when two or more persons share the same height. However, I feel that BMI does not correctly show whether the person is overweight, obese or not. For example, the people with a heavy bone structure always weigh more than those with a thin bone structure, in some cases, the latter can actually be fatter than the former.

The risks for having many diseases, such as cardiovascular and neurological disorders, are more closely linked to the build-up of fat in the abdomen. Some studies linked the risks to the ratio between the waist circumference and BMI. A study by Dr. Cuilin Zhang and coworkers, which was published in the April 1, 2008 issue of *Circulation*, found that women with a larger waist circumference had a much higher risk of cancers, coronary heart disease, and other diseases. (69)

In my opinion, the height-waistline index, by dividing the 200 times of the waist circumference over the height minus 80, may be the best way to reflect the weight situation of the individual. For that purpose, I developed a simple equation:

HWI = (200 X Waist Circumference / Height) -80

We can use either centimeter or inches for both the waist circumference and the height. When we are on our back, the fat in the abdomen should evenly spread out in all directions, unless the abdomen has scar(s) in an unusual situation. The umbilicus (belly button) is the center of the abdomen when the person is flatly on his back. Measure the waist circumference at the level of the umbilicus while the person is lying flatly on his back. The measurement should be the best way to find the amount of fat in the abdomen. A chart is posted online at www.carbohydratescankill.com. for your reference.

Before examining the effectiveness of the Height-Waistline index, I tentatively suggest that a person with an index between 10 and 20 is within the normal range; below 10 is underweight; between 20-30 is overweight; and over 30 is obese.

More Exercise, Vegetables, And Fruits?

Health experts always tell us that if we do more exercise and eat more vegetables and fruits we will lose weight and stay healthy. To judge the credibility of this advice, we must understand the effects of following the advice.

Exercise definitely helps us use more energy that we took in from foods, especially the extra blood glucose from carbohydrates. (160) But, we should stop a moment and think how many calories we can burn up in exercise for 30 minutes. Of course, during the same period, the amount of calories spent in exercise depends on the types of exercise and the person's body weight. The higher the person's body weight and the harder the type of exercise, the more calories will be burned.

For example, for a 30-minute period of exercise, very fast bicycling burns 300 calories for a 130-pound person and 500 calories for a 190-pound person; slow walking spends 75 calories and 110 calories respectively; climbing stairs uses 450 calories and 680 calories respectively. Not all exercise helps burn enough calories to promote much weight loss.

More importantly, the types of foods we eat make a big difference in planning weight loss. A recent study showed that if a person eats carbohydrates after exercise, his body does not burn fat any more than another person who eats carbohydrates and does not exercise at all. Only the person who eats gelatin (protein) after exercise burns body fat. (219). This explains why the construction workers and laborers continue to gain weight, despite the fact that they use lots of energy in their job. Many of them eat plenty of carbohydrate foods. These foods are the buns of the hamburger, potato chips, French fries, juice (sugar-added), and soft drinks sweetened with sugar, and et cetera. This is the same problem for those people who exercise daily yet still cannot flatten their stomachs.

Oh, yes. Exercise also helps us shape up the body to make it look better and build physical capability, especially in the cardiovascular system. But, have we heard that quite a few athletes collapsed in the middle of games or suffered heart attacks? These tragedies help us realize that more exercise alone does not guarantee good health. We still have to eat the right foods for a healthy life.

So, does eating more vegetables and fruits benefit our health? My answer is that it depends on what we eat.

Fructose and starch have the same potential as other carbohydrate foods when they enter the body and become sugars. (100, 132) We should teach ourselves about the carbohydrate content of each kind of vegetable and fruit. We should allow ourselves only a limited daily amount of carbohydrate food. We should not take vegetables and fruits lightly or eat as many of them as we wish! For example, in my personable experiment, a half medium-size persimmon (55 grams) raised my blood glucose more than 35 mg%, for a period of longer than 30 minutes after eating it. I suspect my blood glucose would be higher if I had not had a carbohydrate-restricted breakfast for dampening the persimmon's glycemic index. I used to eat two or three persimmons for one snack. How high was my blood glucose after snack? I do not know, but am very scared to even think about it!

Of course, eating more fiber helps reduce the total glycemic index and glycemic load; it helps increase the volume of the food, and the feeling of fullness of the stomach. Thus, fibers are helpful indirectly in lowering our blood sugar level after eating. In my opinion, the advice, "more exercise, more vegetables, and more fruits benefit our health", is not totally false, but it is somewhat misleading.

Fat And Protein For Satiety

For the last five-plus decades, health officials and medical professional organizations have told us that we should stay away from fats, especially saturated fats, and cholesterol. They told us to eat more carbohydrates instead. During the recent years, they told us to avoid that extra amount of calories. However, they still told us to avoid fats in favor of carbohydrates. But why are more people getting heavier and becoming diabetic? Why are more people suffering from cardiovascular and neurological diseases? Why are more people developing arthritis, cancers, and many rare illnesses? Why are more children getting diseases used to occur mostly in adults?

We did not know much about how protein impacts our body and health. We knew a patient with a bad kidney who suffered uremia (accumulation of urea, which is a product from protein.); we thought he should stay away from protein in his diet. This is probably the reason why the medical doctors and nutritionists have not encouraged us to eat more protein. In fact, we have worried too much about the impacts of fat and protein on our health to really gain their benefits. Not to mention that we need fat for some essential fatty acids, which our body cannot produce. Also, we need some essential amino acids, which our body cannot produce. Amino acid is a basic unit of protein. Our body needs proteins for tissue repairs, replacement, and growth.

By the way, many of us, with a history of gallstones, might likely have preferred carbohydrates to fats, during the years before gallstones became a problem. The gallstone formation was not a result of eating too much fat. Rather, it was a result of eating too little fat and too many carbohydrates. Eating too little fat idles the gallbladder. The gallbladder keeps collecting bile juice from the liver and concentrates the bile juice, making it more likely to form gallstones. Besides, high blood glucose from eating too many carbohydrates increases the level of inflammation in the whole body. Inflammation involves the gallbladder, too. When the person with gallstones eats fat, the gallbladder needs to squeeze the bile juice into the duodenum. Now, the gallbladder cannot do that without pain, because of gallstone or inflammation of the gallbladder.

In my opinion, there are two important key factors in our feeling of hunger. I call them physical and physiological hunger. The first factor is an empty stomach or physical hunger. After the stomach moves all the foods into the intestine in about 3-4 hours after meal, we realize that the stomach is empty. Of course, the time it takes for the stomach to move all the foods into the intestine depends on what we ate for the previous meal. If we ate more fat and protein, the stomach would have taken a longer time to empty itself. However, we do not feel bad if we do not feed ourselves at that time.

The second factor is a low blood glucose level or physiological hunger. The normal fasting blood glucose level is between 70 mg% and 100 mg%. When the blood glucose goes down to the fasting blood sugar level, we feel that we need to re-supply our body with foods. This situation can be worse when our blood sugar level reaches 70 mg% or lower. At that time, the body becomes weak and we break into a cold sweat. We may be about to pass out if we do not eat something like a piece of candy, cookie, or toast, or drink a glass of juice to raise the blood sugar. Medically, this is an episode of hypoglycemia or pre-diabetic.

The physiological hunger is a vicious cycle that we should avoid. After we eat a lot of carbohydrates, the blood glucose goes up. In response to a rising blood glucose level, the pancreas produces and sends out insulin to lower the blood glucose level. Insulin converts glucose into glycogen, which is stored in the muscles and liver for future use; it also converts glucose into fat, which is stored in the fatty tissue in the organs and flesh. Initially, the pancreas produces and releases enough insulin to take care of the blood glucose. When we eat more carbohydrates, this increases the blood glucose level; the pancreas "feels" that it needs to produce and releases more insulin. More insulin from the pancreas lowers blood glucose level much sooner, thus we need more snacks before the next meal. Sooner or later, we will experience an episode of hypoglycemia, when the pancreas produces more insulin than our blood glucose level needs.

The body has a limited storage capacity for glycogen. More carbohydrates in snacks are eventually stored as fatty tissue. We become heavier!

When we choose to eat a limited and smaller amount of carbohydrates, excluding fibers, our blood glucose will rise only slightly. The pancreas will have to produce and release a small amount of insulin, which is just enough to keep our blood glucose level in the normal fasting range. However, the blood glucose level will not go lower as long as the pancreas takes it easy. In the meantime, we eat more fat and protein, which makes our stomach take a longer time to empty. Therefore, we do not feel hungry so soon; at least, we do not experience the physiological hunger. This is the best way to control the diet and lose weight.

Of course, eating more calories than the body needs can make us grow heavier. With the exception of people who have compulsive eating disorders, we should feel a full stomach after eating so much fat and protein. This is why a carbohydrate-restricted diet can help us lose weight better than a low-calorie, low-fat diet.

Alcohol And Health

We have heard so many conflicting reports about the impacts of alcohol on our health. Now and then, reports told us that a "small amount" of red wine everyday benefited our health, especially for our heart. In between the encouraging reports on alcohol, other reports told us about discouraging impacts of alcohol on our health, such as the development of cancers not to mention chronic alcohol abuse and liver cirrhosis. Whom should we believe? In my opinion, we should avoid using alcohol, which is more harmful than not to our health.

Alcohol is an excellent source of energy for the body. It is readily converted into energy (7 kilocalories per gram) and delays the usage of blood glucose. It results in hyperglycemia when we eat carbohydrate foods at the same time. The pancreas will send insulin out to lower the blood glucose. As soon as the alcohol is totally used by the body, there will be a stage of hypoglycemia until new glucose is regenerated from glycogen and other sources. Both hyperglycemia and hypoglycemia are problematic for our health! I included many articles about alcohol in the reading list online for your review.

Carbohydrate-Restricted Diet For Health

The biggest mistake that we have made in dietary recommendation is assuming that carbohydrates were a wonderful and harmless nutrient as a major source of our daily energy. Based on this serious misconception, we figure out the amount of daily requirement of calories for each person, based on the person's gender, body built, and daily activities. We fix the percentage points of the daily requirement of calories from each nutrient in such a terrible way that a person who needs more calories would have to take in a larger amount of carbohydrates. This is so harmful to the person's health.

Based on this wrong concept, we call different diets low or high of each nutrient, such as low or high carbohydrate, as if a fixed percentage of the dietary recommendation was the gold standard. In fact, the actual amount of carbohydrates in a diet for the particular person is the most important to his health. I strongly suggest that we should never again use the terms such as high- or low-carbohydrate diet, high or low fat diet, or, high or low protein diet. Such terms are dangerous, or misleading to say the least.

When we fix the daily requirement of protein, there are four types of diets, namely (1) carbohydrate-rich (more than 150 grams daily) and fat-rich (unrestricted); (2) carbohydrate-rich (more than 150 grams daily) and fat-restricted; (3) carbohydrate-restricted (less than 150 gram daily) and fat-rich (unrestricted)); (4) carbohydrate-restricted (less than 150 gram daily) and fat-restricted.

Among these diets, the carbohydrate-restricted and fat-restricted diets definitely cause weight loss. Because its total calories are less than the daily requirement, this diet is much like starving and, therefore, unhealthy.

The carbohydrate-rich and fat-rich diets provide more calories than the daily requirement. They will cause weight gain and obesity. They also provide a high level of blood glucose after each meal, and are most likely to lead to the onset of diseases, including diabetes mellitus (DM). Because of the high amount of blood glucose after meal, the insulin level increases; in turn, that causes hypoglycemia and hunger for more foods, especially the carbohydrate.

The carbohydrate-rich and fat-restricted diet is likely to provide more calories than the daily requirement. It will cause overweight and obesity, if there is no restriction on calorie. The blood glucose level after each meal increases; in turn, it increases the risk of developing diseases including diabetes mellitus (DM). The increase of the insulin level causes hypoglycemia. Thus, that causes hunger for more foods especially the carbohydrate.

The carbohydrate-restricted and fat-rich diet increases satiety. It offers the individual more self-control in deciding the type and amount of foods he needs. In most cases, the individual does not have the feeling of hunger and takes in fewer calories than the daily requirement. He would begin to lose weight, as long as he does not increase the amount of carbohydrates in his diet. The lower blood glucose level after each meal keeps a low level of inflammation and prevents the body from developing diseases and aging prematurely.

Just look again at my drawing in the first topic of this chapter. A high blood glucose level after eating a meal can cause excessive weight gain or obesity. However, a person who is of normal weight or underweight does not necessarily have a normal blood glucose level after eating a meal.

For example, a person with a normal BMI and "HWI" may favor more carbohydrate foods. He takes in just enough calories for running his daily activities. Therefore, he is expected to neither lose nor gain weight, yet his blood glucose level rises higher after each meal. His blood glucose falls only until his pancreas releases enough insulin to bring the blood glucose level down. During the window between the time it rises and then is disposed, the higher level of blood glucose can cause acute and chronic inflammation. It also can cause blood clot formation, as well as cardiovascular diseases including hypertension, arteriosclerosis and atherosclerosis, glycation and others. When the high level of blood glucose damages the pancreas, he will have diabetes (DM) and a host of other illnesses.

Having understood the impacts of high blood glucose level on our health, we must restrict the daily amount of carbohydrates in our diet. The carbohydrate-restricted diet keeps our blood glucose within the normal range; it is for keeping us healthy not just for weight control. It prevents our body from both developing diseases and aging prematurely.

Everyone Needs Annual Blood Glucose Screening

The level of blood glucose (sugar) is critically tied to our health. The higher blood glucose level decreases many important physiological functions; it increases the seriousness of many diseases.

The current criteria for diagnosis with diabetes mellitus focuses only on the readings of fasting blood glucose (FBG) and the two-hour postprandial blood glucose (two hours after meal, or drinking a solution with 75 grams of glucose in GTT or glucose tolerance test.

The readings of FBG and GTT only give us a reading of the ability of the pancreas to lower the blood glucose. They do not tell us how high the blood glucose level rises after a meal, within the window of two hours, and before the pancreas releases enough insulin to bring down the blood glucose level. In my opinion, the high blood glucose level during the two-hour window is actually most critical to our health in both short and long terms. I liken this practice to an interesting scenario. To determine a person's good behavior, I would ask him to report to my office promptly at 8:00 AM. Then he could go anywhere and do anything he wants, as long as he reports back to my office at 10:00 AM sharp. To recognize his good behavior, should I ignore what he might have done during the window of two hours? In my opinion, I should not ignore that, especially if he has done some terrible things during the two-hour window. Many research results underscored my observation.

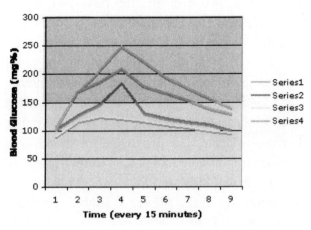

Figure 39. Current Blood Glucose Tests
Do Not Tell US the Whole Story

The above figure represents the two-hour blood glucose levels taken from four people. The readings at the time (1 on the time axis) before eating meal are the fasting blood glucose levels. The readings at the two-hour (9 on the time axis) after eating meal are the blood glucose levels at the two-hour mark. Based on the current diagnostic criteria for diabetes mellitus, none of the people had the disease, but three of them have certain periods of time when their blood glucose levels rise in the gray area, with readings over 150 mg%. The blood glucose level, in the gray area and rising, causes acute inflammation; it attacks and damages our tissues and organs including the pancreas. There are at least three attacks a day. Eventually, the body gets ill.

The case in point is the report by Jee, SH, and coworkers, "Fasting serum glucose level and cancer risk in Korean men and women." Journal of American Medical Association, (JAMA). Volume 293, Number 2, Pages 194-202. January 12, 2005. The report found a positive link between the fast blood sugar (glucose) level and the risks of many cancers in 1,298,385 Koreans. It reported that the diabetic patients had the highest cancer risks. However, there was a "little" inconsistency in evaluating the link between the group with the lowest fasting blood glucose level and their cancer risks. The lowest fasting blood glucose group did not necessarily have blood glucose levels within the "normal" range at all time after meals. Instead, some of the lowest fasting blood glucose group might have had spikes in their blood glucose level after meals that increased their cancer risks.

One interesting facet of Jee's study was that many participants had a normal body mass index (BMI). This again emphasizes that people with a normal or lower BMI can still have a high risk for many diseases. Rather, the blood glucose levels at all time are the most important factor in assessing an individual's risks for diseases.

Having pointed out my observations, it is in everyone's best interest to have his blood glucose checked at least once a year. The check-up should include a series of blood glucose levels for one day or at least for one meal. It should include not only the readings before meal, but also those at intervals of 15-20 minutes for the next two or three hours. (Please see my blood glucose experiments in Chapter 5.) The test will not only help determine the capability of the pancreas in handling the blood glucose. It will also help educate each of us about the relationship between the levels of blood glucose and both the types and the amount of foods we eat. The test will help prevent us from developing diseases and keep us healthy.

After reviewing several research reports, I feel that our blood glucose level should be kept below 150 mg% at all times and the fasting blood glucose level should be under 110 mg%. With the blood glucose levels kept below the suggested limits, our body will have a low inflammation level and a strong immune system. That means lowering the risks of many diseases, such as cardiovascular diseases, Alzheimer's disease, asthma, allergic reactions, infections, chronic obstructive pulmonary disease, cancers, and et cetera.

Critical Dieting For Maternity And Childhood

Studies have shown the links between the newborn's birth defects, such as Down's syndrome, to mothers who have had diabetes (DM) before or during their pregnancies.

The Society of Obstetricians and Gynecologists of Canada published "Teratogenicity Associated With Pre-Existing and Gestational Diabetes," in its SOGC Clinical Practice Guideline, in November 2007 (719). Despite the facts of the flawed criteria for making diagnosis with diabetes (DM), the abnormal level of mother's blood glucose decides the risk of having a baby with malformation and birth defects. The mother's age plays very little role in creating this serious misfortune.

The data shows that more children have been getting heavier during the last decades and that many have been diagnosed with diabetes mellitus. The age, at the time of diagnosis with diabetes mellitus, has been getting younger and younger. Excess weight/obesity and diabetes mellitus are linked to cardiovascular diseases. Thus, we are alarmed that adult diseases have also moved into childhood. Preventing overweight/obesity and diabetes mellitus in childhood has become the top priority in public health.

Studies have pointed out that nutrition during the maternity can influence the health of both the mother and the newborn. Carbohydrate is the most important nutrient that we must address carefully. Carbohydrates are our most common source for supplying energy. It becomes sugars, mainly glucose, when entering our blood circulation. With the help of insulin, the body converts it into glycogen, which is stored in the liver and muscles for future use. Glucagon converts glycogen into glucose when our body needs more energy. Glucose can be used in the production of glycerol and fatty acids, which are stored in the fat tissue. Glucose also produces cholesterol. Cholesterol is used for making steroid hormones, including sex hormones. As explained, high blood glucose level can also be harmful.

In a case of hyperglycemia during pregnancy, we observe not only complications on the pregnant woman, but also the serious effects on the fetus, due to too much blood glucose coming through the placenta from the mother. Studies found that hyperglycemia in pregnancy could cause gene mutation in the fetus. Mutation might be responsible for diabetes (DM) in the newborn. But, I suspect that high blood glucose in the fetus from the mother probably damages the fetus' pancreas, as it does in much the same way that hyperglycemia damages the adult's pancreas. Before birth or shortly after birth, the newborn's pancreas is already wiped out. The newborn will be diabetic in his infancy or early childhood, especially when the infant or toddler is fed with foods rich in carbohydrates.

Physicians always refer "tight glycemic control" to using medications, including insulin, to keep the patient's blood glucose level under control. The most effective glycemic control should include the glycemic content of the diet.

We should recommend a reasonably low-carbohydrate diet for the pregnant woman, to keep the fetal blood glucose within the normal limits. This will prevent new cases of diabetes (DM), or type 1 diabetes, in infancy and childhood from happening in the first place. This will also reduce the risks of congenital malformation including Down's syndrome. Keeping normal blood glucose level at all times during pregnancy should prevent unnecessary obstetric complications. These complications are hypertension, eclampsia, and et cetera.

Children are different from adults in that they have more active factors and hormones for helping their bodies grow.

These factors and hormones help insulin handle more blood sugars for energy and growth at the same time. However, when they are fed with too many carbohydrates, their bodies may become overweight, obese, and ill. This is especially true when we feed children carbohydrate foods with high glycemic indices, such as sugar, candies, and starchy foods. The daily amount of carbohydrates or glycemic load matters very much too.

Just take a look at the foods we feed our children today. Yes, we are trying to avoid "junk foods' or "fast foods." We blame greasy foods from the fast food restaurants for our children's obesity, diabetes mellitus, and other problems. We do not want our children to eat fried foods. But we are doing the right thing for a wrong reason. We should avoid fried foods for glycation between carbohydrate and fat or protein. We should avoid the coating on the chicken skin for frying or baking, not the greasy skin. The glycation products are toxic to the body.

The foods we give our children at home are not much better! Just look at the nutritional facts of the label on the food container or package. How many carbohydrates do we give our children in these foods? Do not forget, milk has lactose and fruits have fructose. Syrup and honey have fructose, too. How about ice cream, cereal, breads, and others? Children should take less of these foods with high glycemic indices, as they grow older.

More studies showed that a ketogenic diet is effective for treating a rising number of diseases. These diseases are uncontrollable epilepsy (conversion or seizure), ADD (attention deficit disorder), autism, allergies, asthma, and others. I expect that many more studies will determine that a ketogenic diet is indeed helpful for these disorders. Not allowing children to eat too much of starchy food, sugar, and sweet fruits should be the first step in preventing these disorders from developing in the first place. The current nutritional guidelines do not emphasize the need to limit the amount of carbohydrates in children's diets. Pediatricians and pediatric nutritionists understand the impacts of carbohydrates on children's health, so they should design a new guideline for pediatric nutrition. We must stop the epidemic of obesity as soon as possible. The epidemic is reaching too many members of the younger generation and shortening their lives.

My Hope

Growing evidence shows the health impacts of dangerous levels of blood glucose. The flawed dietary recommendations are based on the misconception about the roles of carbohydrates in our health; the roles are responsible for an increasing number of preventable illnesses; they are also responsible for our skyrocketing health care costs. Advancements in technology should not only improve the discovery of new diseases, but should also reduce the development of all diseases. In future medical studies on the roles of carbohydrates, fats, and proteins, we should keep two of the three at safe levels during testing. In doing so, we can increase or decrease the amount of one nutrient at a time to see how it affects our health.

Recent studies found that all fats and saturated fats are not necessarily an evil nutrient. However, we are told to stay away from all fats and saturated fats, if possible. (99, 139, 140, 154, 156, 157, 169, 170) This is especially true of cholesterol and eggs. (125, 127, 215, 361)

Inflammation is strongly tied to many diseases. These include cardiovascular diseases, neurodegenerative diseases such as Alzheimer's disease and ALS, as well as cancers. More studies should provide us with a clear picture of how carbohydrate intake is linked to hyperglycemia, high levels of acute and chronic inflammation, and the development of diseases. After recognizing the links, we should be able to effectively prevent many diseases from occurring while treating the ongoing ones.

The cases in point are the recent reports about using ketogenic diets for slowing down the progress of cancers in animal models, for improving the orientation in the animal models with Alzheimer's disease, and for treating the uncontrollable epilepsy in children. We should expect that lowering the amount of carbohydrates in our diet, without forming a ketone body, could produce equivalent results too. These research projects deserve more supports, especially financially, from all of us.

By the way, there is an off-label cancer treatment, insulin potentiation therapy, or IPT. IPT was discovered by Perez Garcia, MD of Mexico in 1926, and continued to this day through his son, Donato Perez Garcia y Bellón, MD, and grandson, Donato Perez Garcia, MD. They gave insulin to their cancer patients along with chemotherapy. They claimed that, in doing so, they could use a very much smaller dosage of chemotherapy agents, yet were able to effectively control the cancers at the same time. No one knows for sure why insulin helped doctors treat cancers with a lower dosage of chemotherapy agents. However, one of the hypotheses on the effectiveness of this approach is that insulin occupies the insulin receptors of the cancer cellular membrane. By doing so, insulin makes the cells more receptive to the chemotherapy agents.

In my opinion, the insulin receptors of the cancer cellular membrane take up insulin, secreted by the patient's pancreas, as much as possible. Then, the cancer cells can take up as much of glucose in the patient's circulation as possible. Glucose is most favorable to the growth and metastasis for cancers. Therefore, there is less insulin and, in turn, less glucose available to the normal cells of the patient. This is probably the reason why cancer patients often lose weight even if they eat as much as they did before they had cancer. Because their cancer cells quickly removed their blood glucose, they were in the situation similar to being on a low-carbohydrate diet.

Interestingly, the available information indicates that the insulin potentiation therapy keeps the level of the cancer patient's blood glucose below 60 mg%, which is hypoglycemia. (http://en.wikipedia.org/wiki/Insulin_potentiation_therapy Wikipedia: Insulin Potentiation Therapy, Claimed Explanatory Molecular Biology). In my opinion, it probably was not the insulin helping the chemotherapy. Rather, the hypoglycemic state was helping the chemotherapy, because the lower blood glucose level limited the source of nutrition for cancer cells to grow and spread.

To prevent the possible risk of insulin shock from the insulin potentiation therapy, the cancer patient may have an alternative. He may require much less chemotherapy agents by keeping his blood glucose level to the normal range, say, below 110 mg%, at all times. Then he must limit the daily amount of carbohydrates, except fibers, and eats more proteins and fats. A lower blood glucose level without hypoglycemia should deny the source of energy for the cancer cells to grow and spread. Of course, limiting the carbohydrates with high glycemic indices should help prevent us from cancer development in the first place.

Hopefully, we may continue to gather the research reports on different aspects of healthy food intake and examine them using a holistic approach. That will help us understand why the diseases develop and how we can effectively prevent and treat them. In the meantime, we should encourage food manufacturers to produce more foods with a low glycemic index. I also hope that there will be more delicious carbohydrate-restricted recipes. These recipes will help people change their dietary habits for health.

There is a label, Nutrition Facts, on the package of almost every food. But the accuracy of the facts is questionable. Hopefully, the FDA of every country and the World Health Organization will jointly set up a universal standard of nutrition facts. The joint standard will help all of us be more aware of the impact of the foods on our health.

My Suggestions For Dieting

For the past two decades, we have become increasingly aware of a growing number of overweight and obese people. We also blame obesity for the increase in cases of cardiovascular disease and diabetes (DM). Weight loss has become a fashionable topic. Many approaches to weight loss are promoted and turned into profitable businesses. However, reports show that most of us fail to sustain our target weight. Many gain back all the weight they lose. Sometimes, they even became heavier than before they started to lose weight. I agree that exercise can help burn up extra calories we pick up from overeating certain foods. Exercise alone, however, is not very effective for losing weight. We should find a way to cut calories at the dinner table, without feeling bad. We have to understand that it is difficult to deny ourselves foods when we are hungry. That means that we need satiety, before we can decide how much we want to eat.

This is the most important advice that I have for all of us: First, we should find a physician who understands how we can lose weight, and sustain the ideal weight. Our physician would give us a physical examination. He would order a baseline laboratory test. If needed, he would order X-rays, electrocardiogram, and vascular studies, before we start dieting for weight loss, more importantly, for health.

To reduce the total calories that we take in, first we have to reduce the amount of carbohydrates to the healthy one. We should eat no more than a range between 125 grams and 175 grams of carbohydrates everyday, depending on sex and height, excluding fibers, for an adult. I would prefer around 100 grams, or less, if at all possible.

For example, an active man needs 2,500 calories a day. Initially, he should cut the daily amount of carbohydrates from 375 grams to 175 grams or lower. His body will get 700 calories from 175 grams of carbohydrates. He may increase the daily amount of fats to 120 grams, which provides 1,180 calories. Also, he takes in 154 grams of protein a day, which provides 620 calories. This approach keeps the same calories initially, while reducing the levels of his blood glucose and insulin in his circulation. He should have no feeling of hunger due to hypoglycemia. Because of the increased amounts of fats and proteins, he will have no feeling of an empty stomach before his next mealtime. Keep it in mind that 154 grams of protein will not provide 620 calories, but only 466 calories. Each gram of protein loses a calorie in converting itself to glucose for energy. Practically, he is taking in less energy food daily, yet he is keeping his regular daily activities.

When he no longer has a feeling of hunger because of hypoglycemia, he will feel a fullness in his stomach with a smaller volume of foods. He is going to lose more weight by burning his fat. I do not recommend a quick weight loss.

Below are my suggestions for dieting:

1. Recruit a physician, before starting a dietary plan. Let your physician give you physical examination, order tests, and help monitor your progress.
2. Buy a good blood pressure monitor and check your blood pressure daily and frequently if necessary. Keep a log of the readings of your blood pressure.
3. Buy a good scale. Weigh yourself every morning after emptying the bladder and bowel, if possible, and before eating your breakfast. Weigh as frequently as you want for the purpose of educating yourself about how foods can affect your weight. Keep a record of the results for review.
4. Learn the nutrition facts on the label of the container of each food.
5. Keep a log of daily calorie intake, and make sure that you do not overeat.
6. Check the online carbohydrate counter for the amount of the foods, which you like, and decide if they are good for you and how much you should

have.

7. Limit the daily amount of carbohydrates, excluding fibers, to the range between 125 grams and 175 grams, depending on your gender, body size (height), and the level of your physical activities. Keep the total daily glycemic load below 150 units, which is equivalent to 150 grams of pure glucose or sugar. If possible, keep the daily amount of carbohydrates to 75-100 grams, if you are on diet to lose weight. Divide the allowed daily amount of carbohydrates into three meals. Never eat too many carbohydrates in one meal to avoid a sudden increase of the level of blood glucose after that meal. Keep it in mind that a sharp increase of blood glucose can result in an acute inflammation, heart attack, stroke, and et cetera.

8. Stay away from sugar, flours, and starchy carbohydrate foods.

9. Avoid carbohydrates that have a higher glycemic index than 50%.

10. Stay away from alcohol and beverages that contain sugar.

11. Eat unsaturated and saturated fats, and proteins as much as needed, including eggs, fish, chicken, pork, beef, and cheese. Avoid animal milk and other milk products, using soy milk instead.

12. Monitor the blood glucose levels at intervals. Prevent unfavorable events as a result of high blood glucose.

13. Eat a good breakfast with more calories, a lunch with more vegetables rich in fiber, and a simple supper with a smaller load of calories and carbohydrates.

14. Collect and make a list of delicious recipes to suit

your taste for a long-term dietary plan.

15. If you wish, you may take multiple vitamin and fish oil everyday.

Most vegetarian diets are unhealthy, because they include a large amount of carbohydrates. The vegetarian diet should decrease or remove sugar, flour products, and starchy foods, while increasing both vegetable fats and protein foods. In fact, such a diet would be much improved by including eggs and cheese for animal fats and proteins, without sacrificing the animal's life.

As emphasized repeatedly, we must realize that dieting is for overall health, not just for weight loss. Those of us who are neither overweight nor obese will still be benefited by a carbohydrate-restricted diet, for good health and longevity. As we grow older, it is a great benefit to eat fewer carbohydrates because the body does not need as much sugar for physiological functions (such as producing cholesterol for hormones) as it does for young people. Eating fewer carbohydrates means avoiding excessive sugar in the body and, in turn, fewer diseases develop.

Never Too Late To Be Healthy Again

It has been six years since I accidentally discovered my dangerous blood pressure readings. I also had cardiovascular symptoms. At first, I failed to help lower my blood pressure with fluid restriction and a no-salt diet; then, I continued my experiments in shaping up my health with a carbohydrate-restricted diet for more than five and a half years. So far, I have reduced my blood pressure to the range of normal readings. In addition, I have regained physical strength and improved the symptoms of the cardiovascular system, as well as other health problems of many years.

I want all of us to know that it is never too late to be healthy again! Studies have supported my statement. (1062, 1063, 1064, 1065, 1066, 1067, 1068, 1069,1072)

Longevity is certainly something most of us want, provided we can experience it in good health. No one wants to have to count pills everyday, as life goes on. Health care and medications are getting very expensive; they are an increasing financial burden. Depleting our financial resource by illness, both before and after our retirement, is something that all of us want to avoid; and we can do that if we want.

The alternative to an unhealthy life is starting now with a healthy dietary plan. The data shows that longevity is not totally a genetic matter. The cases in point are that identical twins who differed in their life spans. The twins had different lifestyles, including diets. A couple of years ago, there was a television news report on a family of four, including the husband, the wife, and two sons. The husband was a chef. All of them were becoming obese over the years. The chef designed a carbohydrate-restricted diet for himself and his family. They were able to reduce their weight to within the normal ranges.

We must understand the choices before us. Are we going to die broke by piling up medical bills while eating "goodies?" Are we going to get healthy again by changing our diet? We must reject the notion that gaining weight is a normal part of the aging process. As we grow older, we should eat fewer calories by cutting down the amount of carbohydrates. In turn, we will keep our blood glucose level to within the normal limits, say under 110 mg%, or at least not over 150 mg%, at all time. This can prevent us from getting new diseases, while helping us to reduce or eliminate symptoms of ongoing diseases. This can spare us from depleting our financial resources for avoidable and unnecessary health care. The savings can provide for a better retirement.

Please keep it in mind: It is never too late to be healthy again!

Acknowledgments

After I failed to bring down my blood pressure with fluid restriction and the no-salt diet during four months in 2002, I questioned the logic behind the relationship of salt and water to blood pressure. I decided to look into research articles for the answer. I found that hyperglycemia has an important role in deciding how our kidneys pick up both water and salt and send them into circulation.

Thanks to the online research from many free databases, I have been able to sample more than 1,200 articles as of this writing. To share the information with someone like you, I have organized a reading list of 1,163 articles online at www.carbohydratescankill.com for your reference.

Also, thanks to the convenience of the Internet I have had the privilege of enlightenment from many world-class medical and nutritional researchers and experts, just to name a few (by the alphabetical order without the individual's title), Breton Barrier, Tom Beer, Jennie Brand-Miller, Todd Brown, Richard Feinman, Stephen Freeland, Mary Gannon, Cheryl Krone, Benoît Lamarche, Ainsley Malone, Tony Nelson, Mark Obrenovich, Walter Rocca, Alessandra Tavani, Paul Thormalley, Maarten Tushuizen, and Jeff Volek. I wish to thank them for sharing their thoughts and personal files with me. Opinions and information from all of their articles have helped me explore the role of carbohydrates in the development of many diseases. They helped me understand the global impact of carbohydrates on our health, an understanding that is very important to the practice of preventive medicine, especially in the areas of prenatal, perinatal, and pediatric care. *My knowledge is far from perfect.* I hope more ongoing researches will help shed additional light on how we should handle carbohydrates in our dietary recommendations especially for our children.

Many thanks to Ms. Jean Morris for her help me in finding the articles that I needed but could not find elsewhere.

I want to thank my children for patiently supporting me for my experiments. I also thank my other family members and friends for their encouragement in getting this book published.

About The Author

ROBERT SU, M.D.

Dr. Su was born in Taipei, Taiwan in 1942. He graduated from School of Pharmacy, former Taipei Medical College in 1965, and received his Bachelor Degree in Pharmacy. He graduated from Nagasaki University School of Medicine in 1971, and received his Medical Doctor degree. He briefly interned at the Department of Surgery, Nagasaki City Hospital, after his graduation from medical school. He completed his internship at Episcopal Hospital in Philadelphia, Pennsylvania in 1972. He finished his residency in anesthesiology at the Medical College of Virginia in Richmond, Virginia in 1974. He has been in private practice in anesthesiology and pain management since 1974. He retired from the practice in anesthesiology in 1997. He continued to practice pain management full time until 2006. He continues to practice pain management, part-time, with acupuncture.

Dr. Su loves to share his experiences with his patients, friends and colleagues like you, and family members. He continues to give presentations on a variety of topics, especially in diets and health. His goal is to prevent the loss of lives from illnesses and violence, through public awareness and education.